AFRICAN

PREDATORS

AFRICAN PREDATORS

Gus Mills and Martin Harvey

Smithsonian Institution Press • Washington D.C.

Published in the United
States of America by the
Smithsonian Institution Press
in association with
Struik Publishers (Pty) Ltd
[a division of New Holland
Publishing (South Africa) (Pty) Ltd]
80 McKenzie Street
Cape Town 8001
South Africa

Library of Congress Cataloging-
in-Publication Data available
Library of Congress Catalog
Card No.: 2001031081

ISBN: 1-56098-096-6

Manufactured in Singapore, not at
government expense
08 07 06 05 04 03 02 01
10 9 8 7 6 5 4 3 2 1

PHOTOGRAPHIC CREDITS

Copyright © 2001 in photographs: **Martin Harvey,** with the exception of
the following:

FLPA/E & D Hosking: p. 32
Paul Funston: p. 116
Gallo Images/Daryl Balfour: p. 30 (left)
Clem Haagner: endpapers, p. 47
Lex Hes: p. 147
J & B Photographers: pp. 83, 128 (top, middle), 135
Gus Mills: pp. 7, 14 (bottom), 19 (right), 46, 68–9, 93 (top),
96–7 (bottom), 97 (right), 98–9, 114 (bottom), 120, 124, 128 (bottom), 129,
134, 136, 138–9, 140–1 (top), 147, 149, 150–1, 155 (top)
NHPA/Christophe Ratier: pp. 39 (bottom), 64, 105;
NHPA/John Shaw: pp. 85 (top), 122
Philip Richardson: p. 108–9

FOREWORD

Africa is the 'cradle of mankind', and we can trace our human ancestry back to this continent through our own genes. This map of who we are and where we come from shows clearly that all humans on the other continents are descended from the small number of tribes that left Africa to colonize the world some 100 000 years ago.

In this sense, all humans are Africans, and the condition of this ancestral continent should be of universal concern. The last significant relics of the megafauna that once strode all the continents are found only in Africa, and many visitors come from all corners of the earth to view the elephants, rhinos, antelopes and the predators.

It is the large predators, usually the big cats, that often grab the most attention, and the thrill of seeing wild lions or leopards in their natural environment is an experience of a lifetime. The wild dog, hyaenas and the jackals also fascinate visitors to Africa. There are in fact 14 large terrestrial predators, meaning those weighing about 10 kg or more, in Africa, and this book provides fascinating facts, figures and insights about them that will enlighten even the most seasoned wildlife enthusiast.

Regrettably, this group of predators is not universally loved in Africa, and several species have suffered from a serious reduction in numbers and range, while some are precariously surviving in protected areas that are like islands in the sea of humanity. Because predators regularly conflict with animal husbandry by humans, and because humans have superior fire-power and technology, the ancient tapestry that wove man and beast together in Africa is unravelling at an ever increasing pace.

The planet is in the grip of the 'sixth extinction', where human activities are causing the rapid loss of species from a variety of ecosystems. Africa, with its rapid population growth and often volatile politics, has the most to lose in terms of wild areas, biodiversity and megafauna. The rapid spread of knowledge about the continent's rich natural heritage is essential in the battle to save it from destruction. The intellectual and moral responsibility to do so rests not only with the people of Africa, but with their close relatives, the people of the rest of the world.

This book does a fine job of spreading the word about predators, their role in ecosystems, and their place in Africa. It should be widely read and disseminated, so that the responsibility of conserving these animals for future generations of humans will be based on an enlightened foundation.

Dr John Ledger
Director, Endangered Wildlife Trust

Endangered Wildlife Trust

Carnivore conservation group

EWT Mission Statement:

The Endangered Wildlife Trust conserves endangered species and ecosystems in Southern Africa by: initiating and funding research and conservation action programmes; preventing species extinctions and maintaining biodiversity; supporting sustainable natural resource management; communicating the principles of sustainable living by education and awareness programmes to the broadest possible constituency, for the benefit of the people of the region.

CCG Mission Statement:

The Carnivore Conservation Group promotes the conservation of carnivores through integrated research on aspects that will help develop and implement sound management strategies.

ACKNOWLEDGEMENTS

I gratefully acknowledge the support of John Ledger, director of the Endangered Wildlife Trust (EWT) and Pat Fletcher of the Carnivore Conservation Group of EWT, and thank them for the interest they took in this book. I thank South African National Parks and EWT for supporting my research on carnivores for over 25 years in the Kgalagadi Transfrontier and Kruger national parks and for making it possible for me to visit many other areas in Africa to observe carnivores. The Tony and Lisette Lewis Foundation made it possible for me to divide my time between S A National Parks and EWT.

To my children Michael and Debbie and especially to my wife Margie, thank you for your support and patience while, instead of attending to domestic duties, I sat in front of the computer writing.

Apart from my own observations, I have relied heavily on the excellent work and writings of many colleagues and friends in the scientific community. As it has not been possible for me to acknowledge them in the text, I have listed the people whose work I used for the various species and, where relevant, the area or region in which they worked. I include three authors whose books I used as general reference works.
Estes, Richard. 1991. *The Behavior Guide to African Mammals.* University of California Press, Berkeley.
Macdonald, David. 1992. *The Velvet Claw.* BBC, London.
Smithers, Reay. 1983. *The Mammals of the Southern African Subregion.* University of Pretoria, Pretoria.

I am conscious of the great contribution made to our knowledge of African predators by those whose names appear here. To those whose names I have not listed, I can only say that I was limited in the number of words I could write and there simply was not room enough to include all the excellent information that is available on this subject.

Lion: Brian Bertram (Serengeti), Peter Bridgeford (Skeleton Coast National Park), Ross English (Kruger), Fritz Eloff (Kalahari), Paul Funston (Kruger, Kalahari), Robert Heinsohn (Serengeti, Ngorongoro), Graham Hemson (Makgadikgadi), Craig Packer (Serengeti), Anne Pusey (Serengeti), Netty Purchase (Zimbabwe), Richard Ruggiero (Gounda-St Floris), George Schaller (Serengeti), Butch Smuts (Kruger), Flip Stander (Etosha), Karl von Orsdel (Queen Elizabeth), Ian Whyte (Kruger).

Leopard: Ted Bailey (Kruger), Christophe Boesch and David Jenny (Tai Forest, Ivory Coast), Koos Bothma (Kalahari), Bob Brain (Paleantology), Hans Grobler (Matopos), Lex Hes (Londolozi), Peter Norton (Western Cape), George Schaller (Serengeti), Flip Stander (Kaudom), Viv Wilson (Matopos).

Cheetah: Tim Caro (Serengeti), Ross English (Kruger), Sarah Durant (Serengeti), John Fanshawe and Clare FitzGibbon (Serengeti), Karen Laurenson (Serengeti), Laurie Marker (Namibia).

Caracal: Koos Bothma (Kalahari), Lucius Moolman (Karoo), Chris and Tilde Stuart (Karoo).

Serval: Jane Bowland (KwaZulu Natal), Aadje Geertsema (Ngorongoro), Rudi van Aarde (North West Province).

Striped Hyaena: Hans Kruuk (Serengeti), L Leakey (northern Kenya), David Macdonald, Rudi van Aarde.

Brown Hyaena: Martyn Gorman (Kalahari), Richard Goss (Namib).

Spotted Hyaena: F Balestra (Malawi), Sue Cooper (Northern Botswana), Ken Drummond (Bubiana), Marion East and Heribert Hofer (Serengeti), Laurence Frank (Masai Mara), Martyn Gorman (Kalahari), Joh Henschel (Kruger), Kay Holekamp and Laura Smale (Masai Mara), Hans Kruuk (Serengeti, Ngorongoro), Netty Purchase (Zimbabwe), Butch Smuts (Kruger).

Aardwolf: Mark Anderson (Northern Cape), Koos Bothma and Jan Nel (Namib), Robby Cooper (South Africa), Hans Kruuk (Serengeti), Philip Richardson (Northern Cape).

Wild Dog: Scott and Nancy Creel (Selous), Ross English (Kruger), John Fanshawe and Clare FitzGibbon (Serengeti), Derek Girman (Kruger), Martyn Gorman (Kruger), Markus Hofmeyr (Madikwe), Darryn Knobel (Kruger), James Malcolm (Serengeti), George Schaller (Serengeti), Mike Somers (Hluhluwe Umfolozi), John Speakman (Kruger), Gus van Dyk (Pilansberg), Joseph van Heerden and Hugo van Lawick (Serengeti).

Ethiopian Wolf: Dada Gotelli, Claudio Sillero.

Jackals: Robert Atkinson (side-striped jackal, Zimbabwe), Dick Estes (black-backed and golden jackals, Serengeti), David Macdonald (golden jackal, Israel), Andrew McKenzie (black-backed jackal, Tuli), Patti Moehlman (side-striped, golden and black-backed jackals, Serengeti), David Rowe-Rowe (black-backed jackal, South Africa).

GUS MILLS

The photographs in this book could not have been taken without the many wonderful people who gave their time, experience and hard work. My thanks go to them and my admiration for their contribution to conservation.

Namibia: The Hanssen family – Donna, Wayne, Lise, and Carla Conradie (AfriCat Foundation); Laurie Marker and Bonnie Schoeman (Cheetah Conservation Fund); the late Nick van der Merwe and his family; Mariet van der Merwe; Marlice van der Merwe for her professionalism and enthusiasm (Harnas Wildlife Foundation); Ingrid Wiesel, for willing support in photographing the shy and elusive brown hyaena (Namib Desert Brown Hyaena Project).

Botswana: Glyn Maude for showing me the brown hyaenas of the Makgadikgadi Pans; Tracy Shapiro and the staff of Mombo Camp, Okavango Delta (Wilderness Safaris).

South Africa: Annie Beckhelling (Cheetah Outreach); J A Strijdom, Brian Jones and, especially, Colin Patrick (Moholoholo Wildlife Rehabilitation Centre); Lente Roode (Hoedspruit Research and Breeding Centre for Endangered Species, Kapama Private Game Reserve); Duncan MacFadyen, for the loan of predator skulls (Transvaal Museum, Pretoria. Department of Mammals).

Ethiopia: Dr Stuart Williams (Ethiopian Wolf Conservation Project, Bale Mountains National Park).

My sincere appreciation goes to Jessica White, Jonathan Harvey, Barbara Mather and Shaen Adey. My thanks also to Dr John Ledger, director of EWT, and to the author of this book, Gus Mills, for his knowledge, insight and advice.

MARTIN HARVEY

CONTENTS

This book is about 14 of the larger African predators belonging to the mammalian Order Carnivora, one of the approximately 20 orders of mammals. In taxonomic terms, hence, they are carnivores (unlike humans or crocodiles or raptors, all of which may be called carnivorous creatures but are not carnivores).

The carnivores described here are those confined to the terrestrial eco-systems of Africa and are species in which the adult females do not usually weigh less than about 10 kg. They belong to three of the ten families generally recognized in the Order Carnivora: Felidae (cats), Hyaenidae (hyaenas) and Canidae (dogs). By and large they prey on the diverse ungulate (hooved) animals of Africa.

Despite the name, however, not all carnivores feed exclusively on meat. Jackals, for instance, survive on a diet of meat and fruits; the aardwolf eats almost nothing but termites; and the giant panda (also a member of the Order Carnivora) lives almost entirely on bamboo shoots. There are over 260 carnivore species on earth and they range in size from the 50 g least weasel to the 2 400 kg southern elephant seal. Most of them are land animals; some like the otter spend much of their lives in water; and about 30 species, the seals and their relatives, are marine, leaving the sea once a year to breed.

What makes an animal, including an omnivorous African predator, a carnivore? The unifying feature of this group is the modification of the fourth premolar tooth in the top jaw and the first molar in the bottom jaw into a set of scissor-like teeth called the carnassials, which are set back in the mouth and are used for shearing through meat.

However, it is not quite as simple as that.

The lion (opposite) is a member of the Felidae, and so is the serval (top left). The spotted hyaena (middle left) is a member of the Hyaenidae and the wild dog (bottom left) of the Canidae, the other two carnivore families that are described in this book.

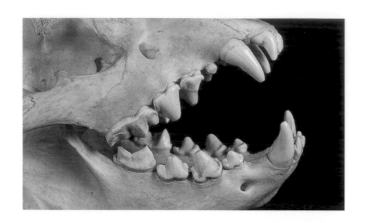

Teeth are the main tools of carnivores, used for killing and feeding. The teeth of the lion (top right) and the spotted hyaena (above) show the lethal canine; those of the spotted hyaena also show the robust molars for crushing bones. The scissor-like carnassial shear – the back teeth on both top and bottom jaws – is clearly visible in the skull of the lion (bottom right) and of the spotted hyaena (top left).

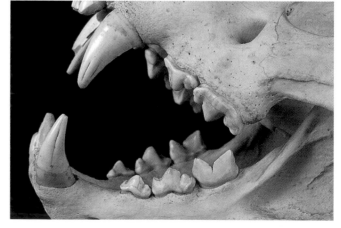

Some carnivores, such as the giant panda and the aardwolf, do not possess carnassials. To qualify for inclusion, it is sufficient that an animal's evolutionary ancestors possessed carnassials and were meat-eating.

The classification and evolutionary history of carnivores are controversial subjects that have been – and still are – the subject of hot debate, discussion and, on occasion, disagreement among scientists. The problem, of course, is that we cannot go back in time. We have to make do with a rather patchy fossil record and try to piece together what really happened. In spite of this, it is remarkable how much of the early history of mammals has been uncovered by palaeontologists, evolutionary biologists and geneticists. The development of accurate methods for dating fossils is of particular importance, and the story is gradually unfolding.

It goes something like this:

At the time of the extinction of the dinosaurs, about 65 million years ago, mammals were small shrew-like creatures. The rapid mass extinction of the dinosaurs – the dominant animals on earth at the time – opened up ecological vacancies or niches, including that of predator, that were soon filled by the mammals. The early mammalian predators were marsupials, the earliest of which was a small, opossum-like creature with a pointed snout and large ears. All shapes and sizes of these early marsupial carnivorous creatures dominated the southern continents for 30 million years.

Except for the canines, which can inflict serious injury in a fight for dominance, the aardwolf's teeth have been reduced to a few peg-like structures. Feeding on termites does not require sharp teeth.

On the northern continents, however, placental mammals were evolving. In placental mammals, the young develop inside the mother's womb, instead of in a pouch after birth. One of these placental mammals was *Cimolestes*, a creature the size of a squirrel that lived on insects. Cimolestes possessed a very important feature, namely a flattening of the cheek teeth, which provided the beginnings of a scissor action. Over several millions of years, these cheek teeth were refined to slice meat in what became the carnassial shear.

Interestingly, the carnassial shear was inherited by two separate groups of mammals: the initially dominant Creodonts and another group that gave rise to the modern Carnivora. In the fossil record from 55 to 35 million years ago, a number of cat-, dog-, bear- and hyaena-like animals are found, some even with sabre teeth, but none of these are true Carnivora. Then the fossil record shows a change: more Carnivora species and fewer and fewer Creodonts. The reason for this replacement of Creodonts by Carnivora is not clear.

One explanation is that the carnassial shear in the Carnivora was situated more to the front of the mouth than in the Creodonts. This meant that the teeth further back in the mouth could still be used for feeding on vegetable matter and other foods. As a result, perhaps, the Carnivora were able to exploit a greater variety of ecological niches than could the Creodonts, who had no teeth behind their carnassial shears and, thus, could eat only meat.

Support for this idea comes from evidence of climatic change during the demise of the Creodonts. The earth became cooler and more seasonal. This may have made prey more scarce, but fruit crops and insects more abundant owing to the seasonal bloom.

Not surprisingly, the early Carnivora, known as miacids, were small and unspectacular, many resembling the genets of today. The major division into dog-like and cat-like Carnivora took place some 55 million years ago. By 7 million years ago, all the modern carnivore families had evolved. Amongst the cat-like Carnivora, the sabre-toothed cats dominated the scene from 26 million to 2 million years ago.

As the Carnivora moved south, they out-competed the marsupial predators mentioned earlier. Today, only a handful of their descendants, such as the Tasmanian

ORDER CARNIVORA: THE CAT DIVISION

Tiger (Felidae)

Spotted hyaena (Hyaenidae)

Yellow mongoose (Herpestidae)

Small spotted genet (Viverridae)

All carnivores belong to one of the 10 families – in two divisions (cat-like and dog-like) – that are generally recognized in the Order Carnivora. They are a diverse and disparate lot, ranging from the 50 g least weasal (Mustelidae) to the 2 400 kg southern elephant seal (Phocidae). Most are land animals, but some spend the major part of their lives in fresh water and others leave the sea only to breed.

devil and the quoll, survive in Australia. Perhaps the best known is the thylacine or Tasmanian wolf, which was exterminated about 70 years ago by bounty hunters.

Most scientists involved in the field of Carnivora classification recognize 10 families in two major divisions: the cat-like families and the dog-like families. The cat-like families are the: Viverridae (civets and genets), Herpestidae (mongooses), Felidae (cats) and Hyaenidae (hyaenas). The dog-like families are the Ursidae (bears), Otariidae (eared seals, namely fur seals and sea lions), Canidae (dogs), Procyonidae (a collection of mainly South American Carnivora including racoons and coatis), Mustelidae (otters, badgers, skunks, weasels and polecats), and Phocidae (true seals, such as elephant seals, monk seals and leopard seals).

ORDER CARNIVORA: THE DOG DIVISION

Polar bears (Ursidae)

Cape fur seals (Otariidae)

Harp seals (Phocidae)

Dingoes (Canidae)

Red panda (Procyonidae)

Honey badger (Mustelidae)

THE CATS
(Felidae)

There are 36 wild cat species on earth, and they are found on all the continents except Australasia and Antarctica. There are 8 species in Africa, 5 of which are included here. The 3 – smaller – African cat species omitted are: the ancestor of the domestic cat, the African wild cat *Felis silvestris*, which is found throughout Africa except in the tropical rain forests; the African golden cat *Profelis aurata*, which inhabits mainly the moist tropical forest zones and about which little is known; and the diminutive black-footed or small-spotted cat *Felis nigrepes*, found only in the arid regions of southern Africa.

LION

Panthera leo

The lion is the largest of the cats. It is a sandy to tawny-coloured animal with white underparts on which the females often retain rosettes and spots characteristic of young lions. Adult males are not only larger than females; they have a mane of long hair on the side of the face and top of the head, extending onto the shoulders. The mane is usually tawny, but in some individuals it may be almost black or, occasionally, reddish or cream coloured.

Of all the cats and, in fact, of all the carnivores in this book, lion and lionness show the greatest sexual dimorphism (the difference between male and female).

Nowhere have more lions been weighed and measured than in South Africa's Kruger National Park. Over 400 of them were handled during an intensive study in the 1970s and subsequently. The adult males among them weighed on average 187,5 kg (maximum 225 kg), and the average height at the shoulder was 120 cm. The average female weighed 124,2 kg, with a shoulder height of about 110 cm.

East African lions tend to be slightly smaller than those in southern Africa, the average weight of East African males being 175 kg and of females 119 kg. Male lions continue growing until they are about 7 years old; females until they are about 10 years old.

In historic times, the area over which lions range has shrunk considerably. Lions became extinct in Europe some

2 000 years ago and disappeared from northern Africa and most of southwest Asia about a century and a half ago. Today, only a relict population of the subspecies *Panthera leo persica* (the Asiatic lion), numbering about 250, survives in India's Gir Forest. At the turn of the 20th Century, the lion was found in suitable habitats throughout Africa south of the Sahara. Here too, it is becoming increasingly rare outside large conservation areas. In the western parts of Africa, numbers are reported to have fallen drastically, and lions are now found mostly in East Africa and in southern Africa. The lion has a wide habitat tolerance and is absent only from extensive desert regions and tropical rain forests. Open woodlands and thick scrub provide the best kind of habitat for lions.

Although their numbers and range in Africa have shrunk over the last hundred years, lions are found in many different habitats, from semi-desert to dense woodland savannah. Viable populations, however, are found only in large protected areas such as the grassy plains of the Serengeti-Masai Mara region in East Africa (left) and the wooded savannah of the Kruger National Park in South Africa (above).

Rosettes, not spots, mark the leopard's coat. A lithe and powerful cat, the leopard is well known for its tree-climbing ability (left). Like all cats, it has prominent whiskers (opposite), indicating the importance of the tactile sense in the Felidae family.

LEOPARD

Panthera pardus

The appearance of the leopard varies from region to region, in keeping with its widespread distribution. Its coat ranges in colour from pale yellow to deep gold or tawny, and is patterned with rosettes, not spots (like the cheetah's). Only its large head, lower limbs and belly are spotted. Its coat colour and patterning are associated in Africa with habitat: savannah leopards tend to be rufous to ochre; desert leopards are pale cream to yellow-brown; rain-forest leopards are golden; and high-mountain leopards are dark. Black leopards, sometimes called black panthers, live mostly in humid forests. They are a colour variant, not a subspecies.

The leopard varies greatly in size across its range, and males are considerably larger than females. In the Kruger National Park, a large male leopard will weigh just over 60 kg and stand about 75 cm at the shoulder, whereas an adult female usually weighs under 40 kg. In the coastal mountains of the Western Cape in South Africa, male leopards are about half the weight of their Kruger counterparts.

Exceptionally large leopards, weighing as much as 90 kg, have been recorded now and again in South Africa.

The leopard is the most widespread of all living wild cats. It is still found over most of Africa (although extinct in Libya and Tunisia), in some of the more remote mountain areas of southwest Asia, across tropical and east Asia, including the islands of Sri Lanka and Java, through to the far eastern reaches of Russia. Although a number of subspecies are recognized, the most recent genetic studies and analyses of skull measurements suggest that only one subspecies occurs in sub-Saharan Africa. The Far Eastern or Amur leopard is the most endangered leopard subspecies, and it is unlikely that there are more than 50 surviving in the border areas of Russia, China and North Korea.

The leopard has a very wide habitat tolerance, and is absent only from the most extensive arid deserts. It can live in rain forests and – in Central Asia – even on mountains up to 5 000 m high, in the domain of the not so closely related snow leopard *Uncia uncia*. Moreover, it adapts fairly successfully to disturbed natural habitats and human settlements.

CHEETAH

Acinonyx jubatus

The cheetah is tall and slender with a small head. It stands about 80 cm at the shoulder and weighs about 40–60 kg. The male is more robust than the female and, generally, about 10 kg heavier. The cheetah's body colour is buff to white, with numerous black spots. The spots extend onto the tail, which is ringed at the tip. The cheetah's face has characteristic 'tear marks' curving down from the inner corners of the eye to each corner of the mouth.

The 'king cheetah' is an aberrant form found in southern Africa (though a skin was recently discovered in Burkina Faso). Its appearance is caused by a recessive gene (similar to that in the tabby form of the domestic cat and the white form of the lion) that coalesces the spots on the upperparts of the body and flanks into bars. It is not a different species or even a subspecies.

The cheetah mainly inhabits drier regions and prefers open areas to forest habitats. In sub-Saharan Africa, its strongholds in the east are Tanzania and Kenya, and, in the south, Namibia, Botswana, Zimbabwe and Zambia. South Africa and Ethiopia also have healthy populations, but the Sahel and Sudanian semi-arid zones in the west of Africa have become badly degraded and support very few cheetahs. Once found over most of north Africa and southern Asia as far as India, there are now only two small cheetah populations in these regions: one in the southern Saharan mountain regions, the other in central Iran, where 50–100 cheetahs hold onto a precarious existence.

The small head, 'tear marks' on the face, long legs and spots distinguish the cheetah. It is buff to white in colour, with black spots – unlike the leopard, the colour of whose coat varies, with habitat, from pale to dark yellow.

Inhabitants of dry regions, cheetahs are generally found in open country (opposite) in sub-Saharan Africa. Only two small cheetah populations survive elsewhere – in the southern Sahara and in Iran.

The king cheetah (below) is a variant – the result of a recessive gene that has caused its spots to run into bars. It is nevertheless a member of the same species as its more conventional spotted cousin – with the tear-marks that distinguish cheetahs even at the tender age of three months (right). Normal-looking parents can produce a 'king' if they both have the recessive gene.

CARACAL

Caracal caracal

The name 'caracal' comes from the Turkish 'karakulak', meaning 'black ear', from the prominent black tufts on the tip of its ears. This feature has also given rise to the common name 'lynx' for this creature, but it is not closely related to these North American and European cats. The caracal is uniformly brick red or tawny brown and stands about 40–45 cm at the shoulder. The male weighs on average from 13–20 kg and the female is about 10 kg.

Like the three (larger) African cats described earlier, the caracal is found outside the continent, ranging across southern Asia as far as central northern India. Like the cheetah, the caracal shows a preference for drier regions, particularly arid habitats with scrub. However, it is also found in evergreen and montane forests, but not in tropical rain forests. It is particularly common in the western parts of South Africa and Namibia, where its numbers are thought to have increased with the extermination of the black-backed jackal.

Although much smaller, the spotted serval (above) might, at a glance, be mistaken for a cheetah. A fellow Felidae, the caracal (opposite), has black tufts on its pointed ears – the source of its name and of its confusion with the lynx.

SERVAL

Leptailurus serval

The serval is a long-legged and long-necked cat. It has a small head and large ears and a pale yellow coat, with black spots on the sides and bars on the neck and shoulder. It stands 60 cm at the shoulder and weighs about 11 kg.

The serval is found in moist savannah regions with tall grass and afro-alpine grasslands, and its distribution is more localized than that of the cats described here. A few relict populations occur in Morocco, northern Tunisia and Algeria, but it is generally found in suitable habitats south of the Sahara. It adapts well to cultivated areas where fields are left fallow and where there is sufficient moisture and shelter.

THE HYAENAS
(Hyaenidae)

There are only four living species of hyaena and, with the exception of the striped hyaena (which is found also in Asia), they are all confined to the African continent.

SPOTTED HYAENA

Crocuta crocuta

The spotted hyaena is the largest hyaena, with a big neck and head, powerful forequarters and less well-developed hindquarters. Its body colour is off-white to brown with dark irregular spots, although some older individuals, particularly females, may lose practically all their spots. An adult stands about 85 cm at the shoulder. The female spotted hyaena is larger than the male – unusual for a carnivore. In southern Africa, females weigh over 70 kg and males about 60 kg. As with lions, both male and female spotted hyaenas are smaller in East Africa, where females weigh about 55 kg and males about 49 kg.

Not only is the female spotted hyaena larger than the male, she also mimics the male's reproductive organs. She has a pseudo scrotum, and her clitoris is large and penis-like. This bizarre organ contains the opening of the uro-genital canal so that the female gives birth through it. Not surprisingly, it is very difficult to tell the sex of a spotted hyaena – no doubt the source of the fallacy that it is hermaphrodite.

The spotted hyaena is the largest of the hyaenas and the second largest carnivore in Africa. A formidable hunter, its large neck and head, strong teeth and powerful forequarters are adaptations for carrying carcasses and crushing bones.

The spotted hyaena has wide habitat tolerance and, until recently, was found throughout sub-Saharan Africa except for the tropical rain forests, the top of afro-alpine mountains and extreme desert areas. Today, like the lion, it is mainly confined to the larger protected areas, reaching its highest densities in East Africa and in southern Africa, in areas where large- to medium-sized prey are abundant. In the west of Africa, its distribution is particularly patchy.

The female spotted hyaena's mimicry of male reproductive organs makes female hyaenas (above) almost indistinguishable from males, except that the female is larger. Once widespread, spotted hyaenas are now largely confined to protected areas in East and southern Africa. The brown hyaena (right) mostly inhabits the drier regions of southern Africa, including the west coast of Namibia.

BROWN HYAENA

Hyaena brunnea

The brown hyaena is a medium-sized dog-like animal with sloping back and a dark brown to black coat except around the neck and shoulders, which are white. It has white stripes on the lower part of both fore and hind legs. Adults weigh about 40 kg, with little difference between the sexes.

Exceptionally large brown hyaenas of around 70 kg have been recorded in areas as wide apart as the Eastern Cape and the Mpumalanga lowveld in South Africa.

The brown hyaena is mainly an inhabitant of the arid southwestern and drier savannah regions of southern Africa. Although its distribution has shrunk during this century, particularly in the southern part of the Western Cape, it is still widespread and, like the leopard, is able to survive close to human habitation. This was strikingly illustrated recently when a brown hyaena was discovered living in Gauteng, between Johannesburg Airport and the city centre.

STRIPED HYAENA

Hyaena hyaena

The striped hyaena is of similar size and shape to the brown hyaena, but has black vertical stripes on a beige to pale grey body, horizontal leg stripes and pointed ears, and a mane along the back which can be held erect. The black throat patch, larger size and more massive head distinguish it from the aardwolf. It stands about 70 cm at the shoulder and its weight varies from about 26–40 kg in different areas. Five subspecies are recognized on the basis of differences in size and pelage, but this has not been substantiated by genetic studies.

The historical distribution of the striped hyaena encompassed the Sahel region of north Africa as far south as

Its massive head and greater size differentiate the striped hyaena (above) from the aardwolf (opposite). The striped hyaena is the only hyaena species found outside of Africa.

Tanzania, as well as the Arabian Peninsula, the Middle East up to the Mediterranean, and southern Asia into Afghanistan and all of India. Today, it is found over much the same area, but in smaller populations, and its distribution is far more patchy, especially in the west and north of Africa, the Middle East, the Caucasus and the Indian subcontinent. Probably the largest continuous population of striped hyaenas that exists today extends from Ethiopia, through Kenya and into Tanzania.

AARDWOLF

Proteles cristatus

The aardwolf is an aberrant hyaena which resembles the striped hyaena but lacks the robust neck and jaws and the large strong teeth of the more typical members of the family. Like the striped hyaena and brown hyaena, it has a mane along the back which can be held erect. Its coat is yellowish-white to rufous, with several vertical black stripes along the body and one or two diagonal stripes across the fore- and hindquarters. There are also several stripes on its legs. The aardwolf stands about 47 cm at the shoulder. Males and females weigh the same, about 10 kg in southern Africa, and is larger (up to 14 kg) – contrary to the norm – in East Africa.

There are two distinct aardwolf populations, which could be regarded as subspecies. The southern population is spread over southern Africa as far as southern parts of Angola, Zambia and Mozambique. It is separated by a 1 500-km strip of moist woodland from the northern population in East Africa, which ranges as far north as the southeast tip of Egypt.

Although much smaller, the aardwolf resembles the striped hyaena – a similarity that has been interpreted as an example of a smaller defenceless creature mimicking a larger and more fierce looking one, for protection.

An exceptional degree of cooperation between hunting lions and individual variation in the roles played by different individuals has been documented on the semi-arid plains of Namibia's Etosha National Park. Here, different lionesses took up different positions during group hunts. Some (the wings) circled the prey, while others (the centres) waited. The stalking wings usually charged the prey and the centres caught them in flight. Each lioness in a pride has its own position in a hunting formation, and hunts in which most lionesses are in position are more likely to be successful than hunts in which they are not.

It is 'common knowledge' that the lioness is the active hunter and adult male lions are virtually parasites living off the females. This is true in open areas like the Serengeti. However, the more wooded Kruger National Park offers a scenario rather different from the sexist stereotype of the male lion. Kruger male lions obtain most of their food by killing it themselves, and male and female lions in Kruger tend to kill different types of prey. Males kill mainly buffalo, but also impala and warthog, whereas females concentrate on zebra and wildebeest.

There are several reasons for these differences. In more wooded areas, male lions do not associate with females as much as they do in open habitats, probably because the extra cover gives the females a better chance to hide their cubs from (potentially infanticidal) strange males. Male lions in less open areas can invest less time and energy protecting their cubs and can look for new females with which to mate. The increased cover in wooded areas of Kruger also makes it easier for the maned, and thus more conspicuous, male to hunt the ready supply of buffalo and impala. Because zebra and wildebeest live in more open areas, only the (less conspicuous) female can hunt them successfully.

The 'king of beasts' – these male lions act true to sexist stereotype by driving off the female (top right) from a gemsbok kill. She tried to worm her way in (middle right), but soon gave up, leaving them to take prime share of the kill. The male lion, however, is not always the chauvinist he is made out to be.

A good example of the feast-or-famine life of a predator comes from a group of two adult lionesses and their seven year-old cubs that I studied in the Kruger National Park. On one occasion, we followed these animals for 14 days continuously. When we started the exercise, their stomachs looked quite empty and we expected them to kill quickly. This was not to be. Their first kill was made only on the third night, when one of the cubs flushed and killed a white-tailed mongoose, but did not even eat it. On the seventh day, another cub snatched up a newborn warthog and dispatched it in less than a minute. On the tenth night, they all fed on the remains of a zebra foal, killed by another pride of lions the night before, but this provided no more than a morsel for each. Their first significant meal came from a four-month-old zebra foal killed on the eleventh night. That was the sum total of their food intake in 14 days and nights! In spite of this, none showed any ill effects.

A few months later, we followed the same group of lions. This time they feasted. During the first eight days and nights, they killed two zebra foals, two adult impala, two adult warthogs, and a porcupine. On the ninth night, they killed a zebra mare and, on the thirteenth, a wildebeest cow. On the last night, when their stomachs were so full we expected them to sleep off the indulgences of the past few nights, they came across a vulnerable buffalo bull and killed him too. When you don't know where or when the next meal is coming from, you have to make the most of your chances!

It is obvious that group hunting gives lionesses a big advantage and enables them to overcome larger prey than could otherwise be hunted. However, long-term studies in the Serengeti have shown that solitary lions have a higher rate of food intake as they do not have to share their food. Group hunting even decreases the rate of food intake among female lions as the spoils have to be shared. The reasons for the social nature of lions appear, thus, to be more complicated.

The ease with which this magnificent male lion drags the 100 kg remains of a kudu cow is an indication of his impressive strength.

THE LEOPARD

An efficient killing machine:

If cats are characterized by stealth and cunning, the leopard is the prototype. The pounce is the major weapon in its hunting arsenal. It is the supreme stalker, investing its cunning and patience in getting as close as possible to its prey, usually not attacking until it is at least within 20 m and preferably within 5 m. Even in the relatively open Kalahari, the leopard is capable of getting really close to its prey before pouncing, using every scrap of available cover to the maximum. Because it relies so much on cover, the leopard prefers to hunt at night.

The short and powerful limbs of the leopard are ideal for leaping and pouncing. Its massive head, powerful jaws and long canines makes it an efficient killing machine, capable of overcoming on its own prey larger than itself. Once it has obtained its meal, it is able to pull it up into a tree to keep it out of harm's way. In addition to its powerful neck, it has strong muscles attached to its large shoulder blades (scapulae) that raise its thorax, improving its ability to scramble up trees with a heavy load. The strength required to hoist a 40-kg springbok carcass, for example, several metres up into a tree, and the apparent ease with which a leopard achieves this, is staggering. The record leopard haul is a young giraffe weighing close on 100 kg. In areas of the Kalahari where trees are scarce, the leopard may drag its kill into a hole.

Watching and waiting. If this stealthy cat (opposite) loses its major advantage – surprise – it is not likely to make a kill. A superb stalker, the leopard uses every bit of cover available to get close to its prey. Its spotted coat helps it to hide in long grass (below) – the kind of camouflage gear that has served many soldiers well.

The leopard engages with the widest range of prey of any of Africa's predators. Among the cats, it is also the arch opportunist. In sub-Saharan Africa, 92 leopard prey species have been recorded, from dung beetles to an adult male eland weighing 900 kg. Although it focuses its attention on the locally abundant herbivores in the 20–80 kg range, it attacks a longer list of other prey species than its close rivals, the cheetah and the wild dog. In the Kruger National Park, warthog piglets, primates and small carnivores extend the list. In the Kalahari, too, I found that small carnivores are a part (about 10 percent) of leopard kills. The most extreme example of a leopard preying on carnivores is from the Ngorongoro Crater, where a leopard brought 11 jackals in three weeks to its camp and ate them. When I was at Nossob Camp in the Kgalakgadi Transfrontier Park, a leopard killed the ranger's wife's miniature Doberman and left it – Godfather style – in the tree outside her office!

In areas with few herbivores in the 20–80 kg range, leopards have adapted well to a different type of food and pattern of availability. At 4 000 m, on Mount Kilimanjaro, leopards live mainly off rodents, and a leopard stranded on an island in the Kariba Dam is reported to have survived on fish. In mountainous areas like the Matopos in Zimbabwe and in the Western Cape, dassies (rock hyraxes) make up the bulk of its diet. In Central African rain forests, it eats duikers and small primates and in the Ivory Coast's Taï National Park, significant leopard predation on chimpanzees has been recorded.

Although the leopard has often been cited as the baboon's main enemy, it has nowhere been found that the baboon is an important prey item for the leopard. Normally, leopards will attack only stragglers and avoid contact with the main baboon troop. Baboons are formidable opponents and will band together to mob a leopard, displaying aggressively and making the most intimidating noises. In an incident in Zimbabwe's Hwange National Park, a troop of baboons mobbed, for over two hours, a leopard that had tried to catch one of their youngsters. At night, baboons sleep on inaccessible cliffs and krantzes or on tall trees from where they can monitor the approach of a leopard. Paleantological studies show that our own ancestors may have had more difficulty keeping leopards at bay.

By the skin of its claws, this dexterous animal (opposite) hangs on to the baboon kill it has dragged and carried up into a tree (top and above), using its impressive climbing skills (see also overleaf). Leopards normally keep their distance from baboons – formidable and intelligent foes that can gang up on the solitary leopard.

THE CHEETAH

Built for speed

The cheetah is built for speed, as its long and powerful legs, wasp-like waist, deep chest (housing large lungs and heart) and small, streamlined head bear testimony. The long tail acts as a counter balance, enabling it to twist and turn at high speed.

There are other less obvious, but critically important, adaptations. The flexible, elongated spine helps to increase stride length. The claws (often referred to as semi-retractable) remain exposed because they lack the skin sheaths of other cats. This enhances traction, rather like the cleats (spikes) on a sprinter's running shoes. The pads on the feet are hard and pointed at the front, and some are grooved like car tyres with longitudinal ridges to improve braking and prevent skidding. The prominent dew claws are used as hooks to trip up or entrap fast-escaping prey. The cheetah's relatively small canines, and consequent reduction in the size of the roots of the upper canines, allow for a larger nasal aperture and, thus, better air intake when sprinting and when holding down and throttling prey after a hard chase.

The cheetah is the fastest creature on earth, and has been clocked at a speed of 112 km per hour. To attain such speed, the cheetah uses tremendous amounts of energy in a very

short burst. This causes a rapid rise in body temperature. If it fails to catch its prey after about 400 m at full speed, it has to abandon the chase and cool down for at least half an hour before trying again.

Not surprisingly, the cheetah prefers to hunt in open terrain. In the Kruger National Park, it frequents the more open areas on the basalt soils in the east. The thick bush on the granites restricts its main hunting advantage. Even on the basalts, cheetahs prefer to hunt along broad drainage lines that have less dense bush.

Should a cheetah encounter prey in an open area, it will stalk very carefully, getting as close as possible to the prey before rushing. When moving through thicker bush areas, the hunt is usually a mere lunge or short chase with little success. Cheetahs are apt to use vantage points such as termite mounds and trees to look for prey, sometimes showing tree-climbing skills that are associated more with that other large spotted cat, the leopard.

The high speed attained by the cheetah when sprinting (below) is a result of several anatomical features, including long and powerful legs, a streamlined head, small waist, deep chest and a long tail.

Designer claws and pads improve the cheetah's traction and braking. The claws (top) act like spikes on a sprinter's shoes. The pads (above) are hard and pointed at the front to prevent skidding and are sometimes grooved like car tyres.

Like the leopard, and in fact all cats, the cheetah often stalks its prey. This is in contrast with the cursorial hunting strategies of hyaenas and dogs, who are not very much concerned with concealing themselves before running at their prey. However, the cheetah relies less than the leopard on the stalk, depending instead on its explosive speed to close the all-important gap between predator and prey.

It is in places like the plains of the Serengeti and the dry river beds of the Kalahari that the true prowess of the cheetah as a sprinter can be witnessed. Both areas afford ideal hunting conditions for the cheetah – cover in the form of bushes and clumps of grass, trees and broken ground, interspersed with large open spaces in which it can attain full speed, and an abundant supply of Thomson's gazelle (in the Serengeti) and springbok (in the Kalahari). The patient observer in these areas may be lucky enough to catch the drama of a cheetah hunt as the cheetah, unlike other large cats, generally hunts during the day.

Once the cheetah has managed to get close enough to its prey or, as often happens, the prey has inadvertently got close to a concealed cheetah, there follows a period of great suspense while the cheetah waits for the right moment. The victim may seem a little uneasy, as if sensing that something is not quite right – or is this just the observer's imagination? Often, a solitary male, or some other lone animal is the target. It continues to feed or looks away – and the cheetah erupts from its hiding place.

Depending on the gap between it and the cheetah, the prey may be able to hold off the cheetah for the next 300–400 m and escape. The cheetah is unable to run at full speed as it closes in on the prey, which begins to twist and turn in a desperate effort to evade the cheetah's final slap with powerful paw and dew claw. These moments are critical because of the rapidly increasing temperature of the hard-working cheetah. It is all over in a few seconds. The prey comes down, often in a heap of dust – or else, the cheetah is forced to stand by and watch it escape.

Impala, springbok and Thomson's gazelle are the most commonly caught prey of the cheetah, although smaller animals like steenbok and hares are also important. Coalitions of male cheetah obtain big meals from animals such as young zebra, wildebeest and gemsbok, and even adult kudu and waterbuck. During the lambing season, cheetahs focus on the easy-to-catch young of prey animals. By producing their young in synchrony on lambing grounds, swamping the predator with prey, the animals the cheetah commonly preys on ensure the survival of a larger proportion of their young than if the young were dropped throughout the year.

Outside the lambing season, cheetahs select less vigilant, solitary animals such as the territorial Thomson's gazelle and springbok males. I found in the southern Kalahari that, although there were one and a half times as many ewes as rams in the population, nearly twice as many springbok rams as ewes were caught by cheetahs. Also, more older animals were killed than would be expected from the actual age distribution in the population. These findings support the often stated, but difficult to prove, assertion that predators select the more expendable individuals and help, thus, to maintain healthy and viable populations.

Totally focused, a cheetah closes in on its prey (opposite), leaping through the air with long and powerful strides. It dispatches its prey quickly (by gripping its throat and cutting off its air supply), setting down to eat (below) before the kill is lost to a competitor.

THE COMBATIVE CARACAL
Size notwithstanding

Adult springbok (40 kg) and mountain reedbuck (30 kg) are not uncommon prey for the relatively small (under 20 kg) caracal. This is impressive enough, but the caracal's record includes a 60 kg adult impala, young kudu and even sitting ostriches. In the Kalahari, I once came across a caracal feeding on a freshly killed large ostrich male that must have weighed close to 100 kg. On closer inspection, all I could find were two small holes at the back of the ostrich's head, indicating that the bird had probably succumbed to the caracal's canines. Whether it was sick or asleep when attacked I was unable to discern, but even so, its demise speaks volumes for the caracal's courage and combative nature.

Leaping into the air like a goalkeeper, a caracal (above) tries to catch a bird with its paws. It is so quick in the air that it has been reported to kill two birds with one swipe. A guineafowl catch (left and opposite) provides this agile cat with a good meal.

The caracal is not only, or even predominantly, a predator of medium-sized mammals; rodents, hares, hyraxes and small antelopes make up the major part of its diet in most areas. In the Kalahari, I have also recorded small carnivores such as black-backed jackal and the African wild cat in its diet, and, in the Karoo region in South Africa, it is known to have eaten polecats, genets, and mongooses. It is also something of a bird specialist and is known to have taken sleeping birds from their perches, including those of such formidable species as the martial eagle, as well as ground-nesting birds, including the heaviest flying bird, the kori bustard. Doves and sandgrouse are reportedly caught when they come to drink. The caracal leaps high into the air, like a goal keeper, pouching a bird with both paws. It is said that the caracal is so fast that it may get two birds with one leap, swiping at the flock with its paws and knocking the birds down. The caracal is the only African cat besides the leopard that, occasionally, takes its kill into a tree.

THE SERVAL

A cat that likes water

The serval is a rodent specialist and is adapted to hunting rodents in well-watered, tall-grass habitats, even in aquatic grasses in shallow water. It is long-legged, not for speed like the cheetah, but in order to see over and into the long grass. It can leap over a metre into the air and pounce anywhere up to four metres on its prey. Its large ears and well-developed auditory-bullae are highly suited to listening for and locating moving rodents, even underground. It has also been seen to operate efficiently in the air, leaping vertically upwards to 'clap' at birds or insects.

Large rats like vlei rats, grass rats and mole rats form the mainstay of the serval's diet, but mice are sometimes an important element and it even eats small birds, reptiles, fish and insects. In the Ngorongoro Crater, frogs are a significant dietary item. It was also at Ngorongoro that a serval was seen to catch its most unusual prey – a feeding flamingo from a small flock at the lake.

In its element in shallow water, the serval (right) is known to eat frogs and, exceptionally, to catch a lakeside flamingo. Wetland conservation is important to servals, which are confined largely to wetland habitats.

A leaper, like the caracal, the long-legged, large-eared serval can jump vertically into the air (opposite) in pursuit of birds and insects, or pounce on a luckless rat (below) from as far away as four metres. It is an important controller of rodents on farm land.

THE HYAENAS

STRIPED HYAENA AND BROWN HYAENA
Adapted for scavenging

The striped hyaena and the brown hyaena have very similar feeding and foraging habits. They are the ultimate mammalian scavengers and share several adaptations to this way of life. The spotted hyaena is in the same class, but it is less dependent on scavenging. All three are equipped with large and powerful teeth and jaws, which make excellent bone crushers. This enables them to eat larger bones than any of the other carnivores and, most importantly, to gain access to the very nutritious marrow locked up inside. Furthermore, they are able to digest the organic matter in bone and obtain energy from it, also something that no other carnivore can do. They have a wonderful sense of smell and are highly mobile, which enables them to cover large distances in their search for carrion.

Carrion from a wide range of vertebrate animal remains forms the bulk of the diet of the scavenging hyaenas. Although antelope are most commonly eaten, other groups such as small carnivores also figure. For example, 20 per cent of the food remains I found at brown hyaena dens in the southern Kalahari was made up of small carnivores. Vertebrate remains are supplemented, to a greater or lesser degree, by other foods such as wild fruit. In the southern Kalahari, the tsama melon and gemsbok cucumber are important for the brown hyaena, not only as a source of moisture, but for their food value. In the Serengeti, the fruits of the green thorn tree (*Balinites*) are highly favoured by the striped hyaena and, in Egypt, it has been known to damage melon fields and date palm plantations in its search for food.

The scavenging hyaenas also eat insects (such as grasshoppers and, especially, beetles) and birds' eggs (ostrich eggs are a favourite). They may, on occasion, kill and eat a small animal. In the southern Kalahari, killed prey makes up less than five per cent of the brown hyaena's diet – small animals such as springhare, springbok lamb, bat-eared fox and korhaan (small bustards). Along the coast of the Namib desert, the brown hyaena feeds predominantly off Cape fur seal pups, but only about three per cent are killed. Here, it also scavenges marine organisms washed up on the shore. In northern Kenya, the striped hyaena is very much dependent on the lifestyle of the local Turkana herdsmen, scavenging around their homesteads, and reportedly killing their goats and sheep, although to what extent has not been properly established.

The hunting of the scavenging hyaenas is unspecialized and opportunistic – a primitive chase-and-grab affair, as Hans Kruuk, the doyen of hyaena researchers, has said with reference to the striped hyaena. Only small animals are hunted, and the hunts are generally unsuccessful. Of the 128 brown hyaena hunts I observed in the southern Kalahari, only 6 hunts were successful.

In solitary pursuit, this striped hyaena prowls along the shores of a lake at dusk, with an eye on the waterbirds in the distance.

Ostrich eggs are prized food. The brown hyaena is easily able to pick up (above) and bite open an ostrich egg (right) – something that spotted hyaenas and even lions cannot do.

The scavenging hyaenas are solitary, predominantly nocturnal, and cover large distances in their search for food. Kalahari brown hyaena move an average of 31 km per night and sometimes over 50 km. A feature of their foraging behaviour is their ability to store excess food for later consumption should they strike a food bonanza. If a hyaena finds a meaty carcass, it will quickly break off a large piece, carry it off some distance and hide it under a bush, in a clump of grass, or even down a hole, before returning to the carcass. If dislodged from the carcass by a larger predator, it can still recover the stored portion. Alternatively, once it has eaten the carcass, it can return to the stored bit for a second helping.

I saw the most striking example of this behaviour one night when I followed an old female brown hyaena in the Kalahari to an unattended ostrich nest with 26 eggs in it. After sniffing at the eggs for a few minutes, she picked one up and carried it off 50 m before dropping it in an open patch. She ran back to the nest and ate 2 eggs, opening them by biting a small hole at the point, and lapped up the contents. She picked up another egg and carried it some 450 m away, carefully placing it in a clump of tall grass. She then returned for another egg, carrying this one off 600 m in a different direction before putting it down under a bush.

Over the next three hours, this old hyaena carried off another 10 eggs in different directions for varying distances, from 150–600 m, hiding them under various bushes, or in grass clumps. Then she ate another egg at the nest, and carried another one off for 600 m, dropping it rather carelessly in a grass clump before lying down. She had eaten 3 ostrich eggs (equivalent to 72 chicken eggs!) and ensured herself a good supply of eggs for the next few weeks.

THE SPOTTED HYAENA

A formidable hunter

Contrary to the facts and despite numerous articles and films, many people still regard the spotted hyaena as a scavenger and somehow less worthy and interesting than the other large carnivores. Old habits and prejudices die hard. The truth is that this misunderstood and much maligned animal is the most successful of the large carnivores in Africa and is equally adept as a hunter and a scavenger.

Like its close relatives, the striped hyaena and the brown hyaena, the spotted hyaena possesses powerful teeth and jaws and is able to crush and digest bones. Such is the efficiency of its digestive system that it can even digest the protein in dried out bone.

When they scavenge from lion kills, spotted hyaenas do not always wait patiently for the larger predators to finish. If the relative numbers are right – three to four hyaenas per lion and no large male lions – the hyaenas may gang up against the lions and force them off the carcass. (This kind of competition between predators will be discussed in a later chapter.)

The spotted hyaena is fast, possesses great stamina and is strong enough to bring down prey as formidable as zebra and gemsbok and, on occasion, even adult buffalo, eland and the young of giraffe, hippopotamus and rhinoceros. Its usual targets are the young of large antelope such as gemsbok and wildebeest. It is an opportunist and eats almost any mammal, bird, fish or reptile. It even runs down flamingos in shallow soda lakes in Kenya and drowns lechwe in the Okavango by swimming after them.

Spotted hyaenas may pick up man-generated organic material, including cooked porridge, offal, and garbage, as well as different kinds of vegetable matter, and such indigestible items as paper and motor car tyres!

Testing a herd of wildebeest (left) three spotted hyaenas are warded off by a bull. Spotted hyaenas are mistakenly labelled cowards because of their method of challenging prey and retreating in order to select a vulnerable animal.

The lethal horns of a gemsbok are excellent defence weapons even when hyaenas (above) make a determined effort to pull a gemsbok down. After a chase of nearly 2 km at speeds of up to 60 km/hr, three spotted hyaenas succeed in killing and disembowelling a large calf (right). Spotted hyaenas are successful in over 70 per cent of hunts after gemsbok calves.

Small gemsbok calves lie hidden in clumps of vegetation to escape predators, but, if discovered, make an easy kill. An ever-alert black-backed jackal is quick to appear on the scene (opposite) to pick up the scraps.

From time to time, people sleeping outside in the bush have been attacked by spotted hyaenas. However, these are rare events, and I know of only a single case where a hyaena dragged a person out of a tent, ripping through a fly screen in the process. There is a notorious case of spotted hyaenas becoming habitual man-eaters. This occurred in the Mlanje region of Malawi during the late 1950s when five to eight people a year were reportedly killed by hyaenas over the period of seven years. These attacks took place during the summer months when the people were sleeping outside. Spotted hyaenas have also been recorded digging up human corpses and, traditionally, many African tribes put corpses out in the bush for them to dispose of.

The proportions in which hunting and scavenging make up the diet of the spotted hyaena vary from area to area, depending on ecological conditions such as the density of lions. In areas where lions and hyaenas are common and present in similar numbers, such as in the Kruger National Park, spotted hyaenas kill less than half their food. In areas where hyaenas outnumber lions, such as the Serengeti and the Ngorongoro Crater, or where both are present at low densities, as in the southern Kalahari, spotted hyaenas kill more for themselves. In the Serengeti, about 50 per cent of their food comes from their own kills. In the southern Kalahari, this figure is about 70 per cent, and in the Ngorongoro Crater, about 90 per cent.

Spotted hyaenas are cursorial hunters and, unlike cats, do not stalk their prey. Individuals often run at moderate speed through a herd of ungulates, apparently surveying the herd. They are cueing in on individual behaviour, looking for young, old or even slightly infirm. Any clues, however subtle, will be picked up and exploited. Once a victim has been selected, the hyaenas chase it until they outrun it or are themselves outrun.

A spotted hyaena gets to grips with an old wildebeest skull. The bone-chewing habits of hyaenas help to explain their conspicuous white calcium-rich droppings.

The tenacity of spotted hyaenas is remarkable. I once tracked three hyaenas that followed a herd of eland for 24 km before finally pulling down a young bull. Of all the experiences I have been privileged to enjoy in over 25 years of carnivore research, there are none more exciting than the spotted hyaena chases after gemsbok across the Kalahari dunes in the moonlight.

Typically, the spotted hyaena hunts singly or in small groups of 2–5 members. The size of the hyaena hunting group depends, to a large extent, on the prey being hunted. Groups of around 11 spotted hyaenas may be found hunting eland and zebra; wildebeest are usually hunted by groups of 3 hyaenas. Hunting group size is not necessarily related to hunting success. Kalahari spotted hyaenas are just as successful when hunting gemsbok calves singly as when they hunt in groups of 6 or so. Of 55 encounters I saw between spotted hyaenas and gemsbok herds with calves, the hyaenas managed to kill a calf in 73 per cent of cases, a hunting success rate surpassed by very few other carnivores of comparable size hunting similar prey. Any study of the impact of predators on prey that does not include the spotted hyaena will give a very false picture. Not only are spotted hyaenas efficient hunters, they are often present in substantial numbers and are highly mobile.

In the Kalahari, a group of spotted hyaenas will frequently travel over 40 km in a night in their quest for food. I have even recorded a distance of 70 km. In the Serengeti, where most of their prey are migratory, spotted hyaenas often leave their territories and go on commuting trips of an average of 40 km one-way to the nearest concentration of prey. Nursing mothers may make 40–50 such trips in a year.

This mobility, as well as their dietary flexibility, makes spotted hyaenas the most successful of the African predators.

THE AARDWOLF

A hyaena that eats insects

The aardwolf is truly a specialist.

This small, aberrant member of the hyaena family feeds almost exclusively on termites. Not only that, the termites are nearly all of the snouted harvester kind belonging to the genus *Trinervitermes*. It is only the species that varies in different localities. In East Africa the preferred species is *T. bettonianus*; in Zimbabwe and Botswana it is *T. rhodesiensis*; and in southern Africa it is *T. trinervoides*. These termites cannot tolerate direct sunlight and are active at night for most of the year, foraging for grass above ground in dense columns. The aardwolf simply licks them off the ground with its large sticky tongue.

However, even the tiny terrmite is not totally powerless. It resorts to chemical warfare when attacked, and the soldiers squirt attackers with noxious terpenes. A foraging column is normally made up mostly of workers. If disturbed, the column's proportion of workers to soldiers changes rapidly, so that the predator starts taking in more and more nasty-tasting soldiers. Eventually, the quantity of terpenes taken in becomes too much and the predator is forced to end the feeding bout. Given their affinity with hyaenas, aardwolves have a high tolerance for terpenes – enough to maintain, once again, the balance between predator and prey.

Winters in many parts of southern Africa may become too cold for *Trinervitermes* to be active. At this time, the aardwolf switches to feeding in the afternoon on the diurnal pigmented harvester termite *Hodotermes mossambicus*. Nevertheless, food is often scarce for the aardwolf in winter in this part of the world, and it may lose up to a quarter of its body weight at this time of year. In the tropical areas of East Africa, the aardwolf enjoys some variation in its diet during the rainy season, when it also eats termites belonging mainly to the genera *Odontotermes* and *Macrotermes*.

Using its large sticky tongue, the aardwolf licks up termites. An aardwolf can consume 250 000 termites in a night.

THE DOGS

THE WILD DOG

A difficult (energy) balancing act

The wild dog is Africa's premier long-distance hunter, rivalled in this method of hunting only by the spotted hyaena. Impala make up the bulk of its diet in most areas in which it has been studied, except on the Serengeti plains. There, Thomson's gazelle tops the bill. Larger species of antelope such as kudu and wildebeest, are important secondary prey in some areas. Smaller prey like steenbok and duiker are probably more often taken by the wild dog than has been recorded, and it is known to eat hares and lizards occasionally, and even eggs. A peculiar food that I have seen wild dogs eating in the Kruger National Park on several occasions is the dried-out shells of millipedes. Perhaps they were getting some trace element from this seemingly unappetising item.

Much has been written about the hunting methods of the wild dog, most of it inaccurate, some of it downright libellous. An extreme example of the latter is R C F Maughn's *Wild Game in Zambezia*, published in 1914. He writes:

> 'Let us consider for a while that abomination – that blot upon the many interesting wild things – the murderous wild dog. It will be an excellent day for African game and its preservation when means can be devised to give practical effect to some well thought out scheme for this unnecessary creature's complete extermination.'

Even the highly respected naturalist and first warden of the Kruger National Park, Colonel James Stevenson-Hamilton, referred to the wild dog as

> '. . . a terrible foe to game with a wasteful method of hunting. There is no other predatory animal in Africa responsible for so much disturbance of game, and in proportion to its numbers, there is none that deals out more wholesale destruction.'

A cursorial hunter: Africa's best long-distance hunter.

In the early days of most national parks and game reserves throughout Africa, wardens and rangers shot wild dogs on sight. They were still shot as recently as the early 1960s in the Kruger National Park. Harry Wolhuter, author of a famous book about the early days in Kruger, *Memories of a Game Ranger*, published in 1948, epitomizes the attitude of rangers towards this species at that time. He writes:

'They used to congregate in packs of twenty to forty; and, as we regarded them as vermin to be reduced without mercy, they afforded us good sport. One day, when out with my dogs, I ran into a pack of about twenty; and, dismounting, I shot three. The dogs pursued the remainder, who presently turned on them and chased them back to where I was standing. I continued to fire at them until I had emptied my magazine hitting one every time since by now they were very close. I then took a spare packet of cartridges from my saddle wallet, re-filled the magazine and shot a couple more, and not many of the pack escaped.'

Why is it that the wild dog has developed such a bad reputation? Is there any truth in the allegations against it? The short answer is 'No'. The wild dog, like any other carnivore, has evolved in balance with its prey as part of an evolutionary struggle. Contrary to what even some of my colleagues believe, it is impossible to detect from the behaviour of prey whether wild dogs are in the vicinity. They do not spread terror or kill indiscriminately. I have seen impala feeding contentedly within 50 m of a wild dog den while the dogs rested in the shade. Hardly the behaviour you would expect near a predator that spreads havoc in its vicinity.

I think that the main reason for the widespread negative attitude towards the wild dog is that it kills by day, whereas most of the other predators kill by night, and that it kills by

Ten dogs in a pack is the optimum. It provides the best balance between the efficiency of hunting (and therefore energy consumed) and the amount of meat available per dog.

disembowelling its prey, not by suffocation or strangulation as the cats do. For some reason, tearing the victim's stomach open to kill it is regarded as less honourable than suffocating it. Animals kill to survive. They do not care how their victim dies, as long as it is stationary and they can eat it.

As a matter of fact, wild dogs usually kill their prey very quickly, faster than lions often do. Once the prey is dead, the pack will tear it apart in no time and get the meat into their stomachs before some larger and stronger predator comes along and chases them away. This could be construed as vicious and murderous, but is merely sensible and necessary.

The members of wild dog packs hunt together. This enables them, for one thing, to bring down larger prey than they could if they hunted alone. The time of hunting is highly predictable, almost without exception at sunrise or sunset. Before setting out for the hunt, the members of the pack take part in a beautiful social ritual. They rally, uttering strange

Pack hunting (above) is invariably the pattern among wild dogs. In wooded areas, however, once the chase has started, the dogs tend to hunt individually.

chittering and squealing sounds, approaching each other, two or three of them often moving shoulder to shoulder, with their ears laid back, tails held horizontal and lips drawn back into a grin. After approaching a dog and making brief muzzle-to-muzzle contact, the group moves off to greet other members of the pack in similar fashion.

I have watched wild dogs hunting impala in the Kruger National Park from the marvellous bird's eye view of a helicopter. Co-operation is not obvious in these hunts, and it looks as if it's every dog and every impala for itself. As the pack comes running up to the herd, the impala scatter in all directions, often using their characteristic 'rocking-horse gait'.

Each dog runs after an impala, but the impala, being faster than the dogs and also having a head start, usually shake the dogs off quickly. However, should a dog manage to keep an impala in sight, it will continue the pursuit. Then the chase is on, and there is normally only one result, for although the impala might be slightly faster than the dog, the dog has superior stamina. Eventually, the impala tires and starts to slow down noticeably, at which point the dog quickly catches up with it and dispatches it, helped, perhaps, by a few others.

Once the impala is killed, the killers often leave it and go back to look for the rest of the pack, leading them to the carcass, a level of altruism not seen in any of the other large carnivores. Altruism extends further when pups are present, The adults stand back and allow the pups to feed first while they keep an eye out for competitors. Because of the largely

independent nature of their hunting methods in woodland savannas, it sometimes happens that a pack may kill more than one impala in one foray. The larger the pack, the better the chance of this happening, and the more important since there are more mouths to feed.

On the open plains of the Serengeti, co-operation among hunting wild dogs is even more apparent. When in pursuit of a gazelle, they will cut across the arc that the gazelle runs and so surround it. As a result of their cursorial method of hunting, wild dogs kill more gazelle in poor condition than cheetahs do. (Long distance runners are more selective than sprinters.) When large wildebeest calves or yearlings are hunted, it takes several dogs to bring down the victim. One or more of the dogs will grab the wildebeest by the nose and neck while the rest tear at the legs and stomach.

When pups like these 3-month olds (above) have to be fed, the adults stand back and let the pups eat first. Ferocious feeders though they are (right), altruism is evident among wild dogs at a kill as among none of the other large carnivores.

During the 1970s, certain large packs of wild dogs in the Serengeti, one in particular, were found to concentrate on zebra. These hunts were quite slow and the first zebra group chased would link up with other zebra, so that as many as 200 zebra would be chased in all. The hunt would end when one zebra fell behind the rest. One of the dogs would catch it by the tail, at which point the other dogs would attack its legs and stomach. While the zebra kicked and struggled, the dogs would be hard pressed to subdue it, but one of them would manage after several lunges to attach itself to the lip, making it stand still and enabling the others to kill it. The 'lip catcher' was normally one of two dogs in the pack, and its presence appears to be an important prerequisite for zebra hunting. These packs continued to hunt zebra for several generations although other large packs in the area did not.

Unlike the other social carnivores featured in this book (the lion and the spotted hyaena), all the members of a wild dog pack, except for small pups, almost always hunt together. The number of kilograms of meat eaten by each dog in a pack follows a U-shaped curve according to the size of the pack. The least meat eaten per dog is in the most common or modal pack size (10 adults). Dogs in smaller packs get more food because there are fewer mouths to feed and those in larger packs get more because they are able to kill larger prey.

Why should wild dogs commonly form packs of this seemingly inefficient size? The reason lies in the delicate balance between the amount of meat eaten and the energy expended in getting it. Hunting success, the size of the prey killed and the probability that the dogs will make more than one kill all increase with the number of hunting adult dogs, while the distance of the chase decreases. Although the normal pack size does not yield the optimum amount of meat for each dog, food intake per dog per kilometre run peaks close to it.

For its size, the wild dog consumes more meat per day than any other carnivore. Although hardly a vicious and indiscriminate killer, a Kruger wild dog puts away an average of 3,5 kg of meat per day, compared with the 3,8 kg eaten by a Kruger spotted hyaena, almost three times its size. This is because of its very high energy requirements. Studies in the Kruger National Park found that a wild dog uses on average 15,3 mj of energy per day. By comparison, a hard-working Border collie sheepdog, which is more or less the same size as a wild dog, uses only 8,2 mj per day, working a 6-hour day. Considering that a wild dog is active for only about 3,5 hours a day, its energy demands are very high. It is not easy for it to maintain a positive energy balance.

The wild dog's high meat intake may explain why it is not a permanent breeding species in arid regions like the southern Kalahari where the food supply is erratic. Unlike lions and other large carnivores that are able to go for prolonged periods without eating, wild dogs may well have to eat much more regularly in order to satisfy their energy-guzzling physiological systems. As will be seen later, this has important consequences for the survival of the wild dog.

THE ETHIOPIAN WOLF

A lone wolf on small pickings

The wolf is usually thought of as a pack hunter of large herbivores: moose, bison and deer. This is certainly true of the grey wolf of North America and Eurasia. Africa's wolf, however, forages alone, for small prey. The wild dog and spotted hyaena have stolen the niche in Africa of cursorial hunter of large prey.

The Ethiopian wolf is endemic to the high-altitude afro-alpine grasslands and heathlands of Ethiopia, in a habitat mostly devoid of antelope. The major herbivores in this unusual ecosystem are diurnal rodents such as the equally endemic, subterranean giant mole rat, which weighs just under 1 kg, and some smaller grass rats. In the harsh Bale mountains, these and other rodents are the Ethiopian wolf's staple food, with the odd egg and small bird thrown in.

Because their prey is diurnal, so are the wolves. It is quite easy to see these critically endangered animals looking for rodents on the high-lying Sanetti plateau in the Bale Mountains National Park. As their prey live mostly underground, or take refuge

A watchful grass rat – favoured prey of Ethiopian wolves. Grass rats and other rodents are the wolves' staple food.

underground when threatened, Ethiopian wolves commonly dig out prey, especially the highly favoured mole rat, leaving behind conspicuous mounds of soil. Prey that is encountered above ground is stalked before being rushed and pounced upon. When it hits a rich food patch, the Ethiopian wolf characteristically continues to hunt without wasting time eating, preferring to cache the food for later consumption once the bonanza has ended.

Occasionally, Ethiopian wolves form small cooperative hunting parties to attack prey such as young lambs and hares or, if available, antelope, falling back on what might be an ancestral hunting strategy. They might also scavenge the carcass of a domestic animal should the opportunity arise. The local name for the Ethiopian wolf is *jedallah farda* which translates as 'the horse's jackal' and refers to its reported habit of following mares and cows that are about to give birth, in the hope of getting the afterbirth. The wolves also hunt among herds of cattle, using the herd as a mobile hide from which to ambush rodents at their holes – a perfect example of carnivore resourcefulness and wolfish deception.

A lone forager, unlike its North American and Eurasian counterparts, the Ethiopian wolf (opposite bottom) is accustomed to sniffing out and digging for its rodent prey. A foraging wolf generally does not take time out to eat (right), preferring to continue hunting and to cache food for later consumption.

THE JACKALS

The supreme opportunists

All three species of jackal eat almost anything: fruit, invertebrates, reptiles, amphibians, birds, all kinds of killed mammals from rodents to impala and almost any type of carrion. There seems to be very little difference in the diet of each of the three species, and their diets appear to be influenced mainly by habitat and the availability of food. In East Africa, the golden jackal is found on the more open plains and desert areas, the black-backed jackal in taller grass and open woodland, and the side-striped jackal in broad-leafed, deciduous woodland.

In the Serengeti woodland regions, the black-backed jackal lives off a species of small diurnal grass rat. However, pairs of black-backed jackals (and of golden jackals) also hunt gazelle and other small antelope. In southern Africa, the black-backed jackal spreads into the arid regions and is found in large numbers even on the coast of the Namib Desert, where beached seabirds, marine mammals and even fish washed up on the shore make up its diet. In farming areas, the black-backed jackal has the reputation of a sheep killer – not without some justification. However, the results of one study showed that sheep losses per month to jackals were no more than 0,05 per cent of the sheep population.

A black-backed jackal in furious – and ultimately successful – pursuit of doves at an Etosha waterhole. All three species of jackal eat almost anything they can get, from fruit to fish.

At the carcass of a cheetah killed by lions (overleaf), two jackals struggle for dominance. Jackals bury what they cannot eat, depriving other scavengers of a meal.

Ever the opportunist, a golden jackal (left) picks up scraps from under the nose of feeding lions on the Serengeti plains. For its part, a young black-backed jackal (below and opposite) finds a beetle an object of great interest and a food item to be investigated thoroughly.

The side-striped jackal is the only one of the three that lives in the Guinea savannah and the belt of miombo woodland stretching across central Africa. It is found on the edge of the equatorial forest and in areas where forest and woodland have been changed into fields. The side-striped jackal is reputed to be less of a hunter than the other two species of jackal and more omnivorous. In Zimbabwe it was found that although the side-striped jackal prefers animal prey (which has a higher energy value), its search for food is aimed largely at fruit. Small mammals are, apparently, only taken opportunistically as and when they are encountered. The side-striped jackal is the only jackal found on the Liuwa plains in Zambia, a habitat that appears to be more suitable for the black-backed jackal. However, an abundant supply of shrub water berries (*Syzigium* species) apparently favours this more frugivorous jackal.

One of the better examples of jackal opportunism comes from Botswana's Tuli Game Reserve, where black-backed jackals regularly hunt old or otherwise compromised impala. Singly or in pairs, the jackals test the impala by running at them and then stopping to watch while the animals run off.

Should they spot a vulnerable impala, the jackals corner it in thick bush, where up to 12 of them have been seen quickly dispatching the animal. Similarly, black-backed jackals in the southern Kalahari prey on the weakness of springbok during the rut, when the springbok males are prone to injury.

THE SOCIAL LIFE OF HUNTERS

passing on the genes

The behavioural repertoire of animals – whatever their ingenuity – is constrained, like their appearance, by ancestral history. Dogs, for example, are usually monogamous and cats tend to be solitary, although (as we shall see) there are exceptions to this rule. However, the feeding habits of animals and the way their food is distributed allow for some fascinating behavioural flexibility.

Procuring live prey is a far more challenging business than procuring herbage. Carnivores tend to be more intelligent than herbivores, and often have more intricate social systems and behavioural patterns that involve co-operation in obtaining food and in defending territories and raising young. The social systems of the carnivores in this book vary from those of the solitary leopard, caracal and serval to those of the less solitary cheetah and the secretly social brown hyaena, striped hyaena and Ethiopian wolf. They extend to the monogamous jackal and aardwolf, and the social lion, spotted hyaena and wild dog.

Animals, ultimately, are slaves to their genes, the units of inheritance that combine to make us what we are. We inherit our genes from our parents and share them with siblings, half-siblings, first cousins and so on. Passing on its genes to future generations is an important element in the life of an animal. Land tenure systems, social groupings, mating behaviour, ways of rearing young, and communication systems all combine in different ways in different species to determine which genes are passed on through evolution and natural selection. The patterns of behaviour that result make for some weird and wonderful stories.

Two concepts may cause confusion: home range and territory. The area in which an animal lives and carries out the functions required for reproduction and survival is its home range; if this area is exclusive to an individual or group (or nearly so) and is marked and defended against others of the species, it is a territory.

The goal of social behaviour is the passing on of genes – whether an animal is solitary like the caracal (top left) or highly social like spotted hyaenas and wild dogs (middle and bottom left) and lions (opposite).

SOLITARY

Leopard, caracal and serval

The leopard, caracal and serval fit the normal cat model in that they lead essentially solitary lives. The only relationships of any length that develop are between a mother and her latest litter of cubs.

LEOPARD

The leopard epitomizes the solitary cat.

In an area inhabited by leopards, there are three superimposed layers of home ranges. First, there is a mosaic of exclusive to slightly-overlapping adult female ranges, the most overlap occurring between the home ranges of mothers and their grown-up daughters. The size of the home range depends on the richness of the habitat in terms of food, and may be as small as $10km^2$ (in the Serengeti), or as large as $200 \ km^2$ (in Kaudom Game Reserve, northern Namibia). It is probably determined by the minimum amount of prey needed by the female and her offspring during the leanest time of the year.

Rearing up to three cubs at once is challenging and takes time (18–22 months); almost half of the cubs (40–50 per cent) die in the first year. Female leopards seldom meet up with each other. They inform each other of their movements through urine scent marks and the characteristic vocalization called rasping, which sounds like someone sawing wood, most often heard in the early evening. By my interpretation, the home range of a female leopard is also a territory.

Overlying the female territory mosaic, is a layer of much larger and more exclusive adult male territories. The maximum size of a male leopard territory depends as much on the size of the female territories as on food supply. In rich areas, where female territories are small, a male will be able to include several females in his territory and still be able to defend the area against competing males. Where the female's range is large, males have to be content with fewer females. In some really poor-quality, habitats, male and female territories may be the same size.

Resident males continually patrol their territories, in the same way as females do, to advertise their presence to other males and to look for females on heat, as well as for opportunities to expand their territory to encompass more females. Because leopards are not seasonal breeders, males must make regular checks on the females in their territory. Strength, aggressiveness and mobility may also influence the maximum size of a male's territory. In this way the 'best' males are most likely to mate with the females and pass on their genes.

The third layer of home ranges is occupied by what are termed floaters. These are usually sub-adults or young adults of either sex that have dispersed from their natal or birth territories and are looking to establish their own home turf. The mortality rate for sub-adult leopards is high: nearly twice that of adult leopards. Food may not be easy to obtain as

hunting skills are still being honed, and resident leopards, which hunt in the best areas, may be aggressive towards the young intruders.

Male leopards usually disperse further than females. It is important that one of the sexes does this as it helps to prevent inbreeding, which is widely believed to be detrimental to a population. By three years of age, a leopard should have established its own territory and be ready to reproduce.

The size of the territory this male leopard (below) has to patrol depends on the distribution of females and the size of their territories. Strength and dexterity (opposite) are also important in defending a territory and, hence, in determining success in mating and, ultimately, reproducing.

CARACAL

The little that is known about the caracal suggests a very similar type of social system to that of the leopard. Radio-tracking studies in the Karoo region of South Africa record slightly overlapping home ranges of 4–6,5 km² for females in the Mountain Zebra National Park and 12–27 km² in the Robertson area to the east. The males appear to range over larger areas, 30 km² and 50 km², respectively, in the two areas, while a radio-collared male in the Kalahari was found to range over the vast expanse of 300 km².

Caracal litters vary from one to four, and caracals first reproduce at about one and a half years of age. A female usually produces one litter per year. Lone cats, female caracals seldom meet up with other females other than their own cubs.

Usually seen alone, the caracal (above) is like the leopard in its solitary pattern of existence. Male and female (below) come together only to mate.

SERVAL

Although its social system is not well known, the serval may reveal the first small step away from the solitariness of the leopard and caracal.

Studies in the Ngorongoro Crater indicate that servals of the same and opposite sex have overlapping home ranges, although they normally avoid each other. A male and female ranged over an area of 30 km² parts of which were used by five other servals. In another area, bordering the Serengeti plains, two male servals had overlapping hunting ranges, although they were never seen closer together than 300 m. In a South African study, the 2–3 km² home ranges of two males overlapped considerably.

As with the leopard, female offspring are tolerated better than male. Male offspring are forced by the mother and resident male to leave when they are six to eight months old. Litters vary from one to three cubs. The serval reaches sexual maturity between one and a half and two years of age.

Serval pairs (below) moving or hunting together are most likely mother and daughter. Male offspring, like this handsome fellow (right), are forced to leave home at an early age.

SLIGHTLY SOCIAL

CHEETAH

If the serval has taken the first small step, among the cats, away from a solitary existence, the coalitions formed by male cheetahs is the next. The female is solitary, associating with males only to mate. Cheetah litters tend to be larger than those of the other large cats, with anything from one to eight cubs (normally three to five). They stay with the mother until they are 18 months of age or older. By this time they are as large as she is, leading to frequent reports that cheetahs live in mixed-sex groups. Two adult breeding female cheetahs have never been known to live together. In the Serengeti, where cheetahs have been studied over many years, contact

More than half of all male cheetahs form life-long partnerships of two and, in some cases, three members. These small groups are better able than solitary males to defend a territory and to gain access to females.

between adult females was seen in approximately 1 in 500 sightings and never lasted longer than a few hours.

On the two occasions that I was able to witness the breakaway of a grown litter of cubs from their mother, I was surprised at how abrupt it was. They were literally there one day and gone the next, and I never saw them come back to her. The sibling group usually stays together for several months until about the time the females come into season or oestrus for the first time .

Then the females break away. to begin the cycle of birth and the hard work of raising offspring unassisted. Why does the male cheetah, or for that matter the male leopard, caracal or serval, not assist with feeding its young? It would seem that in these species the female does not need the male's help to catch enough prey. As hunting success is probably not improved much by the male, his presence would lead only to increased competition with his offspring for food. Better that he moves out of the way and gets on with the job of defending his territory and looking for receptive females through whom to pass on his genes.

Six weeks old and still with mother (below), these cubs will remain with their single parent until they are about 18 months of age and almost as large as she is.

Territorial males continually reinforce ownership of their territories by scent-marking (above) all over the territory, not merely on the boundaries.

The fate of the males from the sibling group is often quite different from that of the females. Only about 40 per cent emulate the females and take up a solitary existence. The rest form life-long coalitions with other males. About 40 per cent of male cheetahs form partnerships of two, and about 20 per cent join in groups of three. Although most males in a coalition are brothers, cases of unrelated males joining together have also been documented. Coalitions – and some solitary male cheetahs – are territorial, scent-marking their territories and defending them against intruding males. Others, including most 'singletons,' are floaters that drift around over a large home range.

The members of a coalition are inseparable. I once studied a coalition of three males, whom I dubbed 'The Three Musketeers', for four years in the Kruger National Park. Almost every time I tracked them, they were together. On the very few occasions when one was missing, the others showed obvious concern, repeatedly making their strange bird-like chirp call until they were reunited. This call carries over a distance of up to 2 km – remarkable for such a soft-sounding noise.

One day, I found two of the Three Musketeers repeatedly calling and looking around for the missing member of the group. About 400 m away, a large group of vultures was sitting in a tree. After a while I drove up to the tree and found the missing male feeding on a kudu calf. I could hear the others calling, and so no doubt could he, but he remained silent, feeding on the kudu. About two hours later, he finally left the kill to the patiently waiting vultures and, with distended belly, joined his partners who were now resting close by. It certainly doesn't say much for the intelligence of the cheetah!

The Three Musketeers. Members of a male cheetah coalition are often brothers and tend to stick together all their lives. One of this trio pays a visit to a 'play tree' (opposite) – the name for certain large trees used by territorial males as scent-marking places. These trees are visited by females when in heat and looking for a mate.

On the Serengeti plains, female cheetahs have vast, overlapping and undefended home ranges of 800–900km^2 that stretch across the plains from northwest to southeast. These cheetahs are, in fact, migratory. Every year, they and their cubs follow the migration of their main prey, Thomson's gazelle, from the short grass plains in the southeast during the rains to the long-grass plains and woodlands in the northwest in the dry season.

In contrast, the territorial males in the Serengeti live on small territories of about 40 km^2. These territories are located around eight different centres, where successive groups of resident males settle. The common feature of these eight centres appears to be a combination of adequate cover for stalking prey and sheltering from enemies and a good supply of gazelle and other prey.

Like the females, floating males in the Serengeti tend to follow migrating gazelle. The floaters are often in poor condition, show signs of physiological stress and are less relaxed than territorial males. They also do not live as long. It is clearly better to be a resident male, and defend a territory, than to be a floater. Because male cheetah territories are in the best habitat, passing females focus on them. Territorial males, thus, have the best chance of finding receptive females and the most mating opportunities.

Combining with one or two other males to maintain a territory obviously enhances territorial ownership. The drawback is that mating opportunities have to be shared with coalition partners. These are usually close relatives such as a brother, however, with a half share in the same gene pool. Thus, even if a male's genes are not passed on directly, some are passed on

indirectly through a relative. This is known as kin selection. Of course not all the females are on all the male territories all the time, so the floaters do get some, if much fewer, mating opportunities.

In the Kruger National Park, we find a very different cheetah land tenure system. Unlike the gazelle on the plains of the Serengeti, impala (the cheetah's main prey in Kruger) are sedentary and evenly spread. In the best habitat for cheetahs in Kruger, female and male cheetahs have overlapping home ranges or territories of about 185 km² each. The Three Musketeers' territory contained almost the whole of two females' home ranges and, partially, that of another female. As they were the only male cheetahs in the area, they

enjoyed mating rights with these females. When two young males tried to come into the territory, they were quickly forced out into far less suitable habitat on the periphery, where they were unable to survive.

As with leopards, cheetah male dispersion is strongly influenced by the way in which the females are dispersed, which in turn is influenced by the dispersal of prey. The end result in the Serengeti is that territorial males encounter more females than do floaters. The differences between social organization on the Serengeti plains and in the Kruger National Park is an example of behavioural flexibility in carnivores, and of how ecological conditions, particularly food distribution, influence patterns of social organization.

SECRETLY SOCIAL

Brown hyaena, striped hyaena
and Ethiopian wolf

One could be forgiven for assuming that the brown hyaena, the striped hyaena and the Ethiopian wolf fall into the same category of social organization as the leopard, the caracal and the serval. Most sightings of these three species are of solitary individuals. Detailed observation has revealed, however, that their behaviour is much more social than it appears to be.

BROWN HYAENA AND STRIPED HYAENA

The brown hyaena takes precedence in this section as its social organization has been much better studied than that of the striped hyaena. In the absence of better knowledge of the striped hyaena, it is assumed that in their social behaviour, as in feeding habits, the two species are similar.

In the course of a six-year study in the Kalahari, I followed brown hyaenas on more than 200 occasions, covering over 3 000 km and (with a San tracker) over 1 200 km of their tracks in the sand. In all this time, I observed two brown hyaenas moving together on only five occasions, and the furthest they moved together was little more than 3 km. However, by spending many nights at brown hyaena dens, I found that some of dens were visited by brown hyaenas other than the mother and her cubs.

These visits were not just courtesy calls. When a brown hyaena visited a den, it usually brought food for the cubs. When food was abundant, these strictly solitary foragers would live in clans of up to 14 hyaenas, almost all of them close relatives, who would congregate at a large carcass and who co-operated in raising the young. Although they foraged alone, clan members shared a common territory – in the Kalahari somewhere in the region of 300 km².

Unlike cats, whose young are kept in a lair such as a clump of bush or thick grass, hyaenas and dogs keep their young in dens, which are usually a hole or series of holes in the ground, sometimes a cave. Although a den may have a large entrance, it quickly narrows into a tunnel too small for an adult or other large animal to enter. The den provides good protection for the cubs against predators during the long periods when adults are away from the den.

Hyaenas enjoy a much longer denning period than the young of cats, who move around with their mothers from about six weeks of age. For the first three months of their lives, the cubs (in litters of one to four) are visited by their mothers and nursed at sunrise and at sunset. As they grow, their mother will visit them only once a night, and supplement their milk diet by meat that she and other clan members carry back to the den. As the cubs grow, the mother's visits slowly tail off, but the cubs are weaned only when they are about a year old. From the tender age of 9 months, they begin to go on foraging sorties from the den on their own, increasingly so as they become more adept at finding food and the contribution of the adults diminishes. At about 15 months of age, they are independent and leave the den.

When they grow, young hyaenas stay on in the territory. The females may be able to breed; the males remain non-breeding helpers, feeding their mothers', sisters', or nieces' cubs, and thus passing on their genes through kin selection.

Two large brown hyaena cubs (opposite top) stand outside a den in the Namib Naukluft National Park. The denning period is the most socially active period in the life of a brown hyaena. Once it grows, it will spend almost all its time alone. Feeding together (left) is a rare occurrence as most of the food that is found provides a meal only for one. At a carcass (above), brown hyaenas characteristically feed one at a time. Each takes a turn, grabbing a piece and carrying it off before the next gets a chance.

In fierce play (above left), brown hyaena cubs of different ages tangle at the den. In interchange of another kind, an adult brown hyaena (above right) pastes onto a grass stalk. On average, a brown hyaena deposits 80 pastings in one night – part of a chemical information network that links it with others of its kind.

Alternatively, both males and females may leave the home or natal territory. The males may take up a nomadic existence, moving extensively, mating with females if and when the opportunity presents itself, and then continuing their travelling existence without investing an ounce of energy in their offspring. During my Kalahari study, I found that one-third of adult males were nomadic. The problem with being a nomad, however, is that females on heat are few and far between.

Another option for a male is to emigrate to another clan, mate with the females in the clan and then take on parental duties by helping to feed the cubs. The challenge here is to ensure that your females remain faithful and do not mate with a nomad. How and why males choose a particular strategy remains uncertain. I think that clans with more than one breeding female are more likely to have a permanent breeding male. A female hyaena, however, cannot breed successfully in a nomadic existence.

Female offspring, thus, must establish their own territories, or choose the option of staying at home and breeding in the same territory as their mother. When females stay at home, they share a den and may even suckle each other's cubs. This is fine when food is abundant, but in times of scarcity the matriarch has breeding priority. At such times, young females may be evicted from the clan.

Because brown hyaenas spend so little time in each other's company, communication through visual and vocal means is limited. The most striking display is the raising of the mane along the neck and back, which seems to occur in most conflict situations and makes the animal look bigger than it really is. Territorial fights are usually ritualized neck-biting bouts between two animals of the same sex (animals of opposite sex do not appear to be aggressive towards each other), accompanied by loud yelling and growling by the submissive animal. The brown hyaena has no long-distance call.

Chemical communication, however, is incredibly well advanced. Each territory is well covered with latrines, at which are left prominent white scats, and with chemical secretions from the hyaena's anal glands. These secretions are deposited onto grass stalks in an elaborate ritual called pasting. When pasting, the brown hyaena deliberately steps over a grass stalk, places it between its hind legs and, crouching slightly, inverts its large anal pouch, like a time lapse of a flower opening, and deposits a blob of white paste on the stalk. As the pouch closes, a thin layer of black paste is wiped onto the stalk. The smell of the white paste lasts over 30 days, whereas that of the black smear disappears in a few hours.

Behavioural experiments and chemical analyses of the pastings have shown that each individual has its own smell and that other brown hyaena can identify this. An individual pastes on average 2,6 times per km travelled. Pastings are so well distributed, that at any one time a territory may have as many as 15 000 active pastings. An individual is within 500 m of an active pasting 99 per cent of the time and within 250 m of a pasting 75 per cent of the time. To a brown hyaena, the territory must reek of its occupants! Because our own sense of smell is so unimpressive, it is difficult to assess all the uses of scent marking. One

function is territorial demarcation. Because territories are so large, it would take a great deal of energy and might well be impossible for a single brown hyaena to adequately demarcate its boundaries. Unless the scent posts were close together, an intruder would easily slip through. Peppering the territory with scent marks means that even if the intruder crosses the boundary it is soon warned.

I have no doubt that brown hyaenas also use pasting to communicate with their fellow clan members. The thin black paste may inform other clan members that a hyaena has recently been foraging in that area and so help to space the hyaenas evenly through the territory. The long-lasting white paste reinforces membership in the clan. For a species that is solitary yet social, leaving messages that can be read later is a good alternative to direct communication.

At 10 weeks of age, a brown hyaena cub can expect to be visited by its mother twice a day – at sunrise and at sunset – to be nursed.

ETHIOPIAN WOLF

Like the scavenging hyaenas, the Ethiopian wolf forages for food on its own and it is mainly at the dens during the breeding season that its social nature is revealed. All pack members guard the den and help to feed the pups by carrying rodents to the den or by the more characteristic method – for a dog – of regurgitation. The members of the pack also come together at night to sleep. They meet at dusk, dawn and even noon, to greet each other and to conduct territorial border patrols, which are usually led by the dominant female. These involve scent marking by way of another characteristic feature of dog behaviour – raised leg urination – and by scratching the soil with the fore feet and defecating on rocks and mounds and other conspicuous sites, as well as howling.

In the best Ethiopian wolf habitat, a pack may consist of up to 13 adults, with an average of 2,6 females to every male and an average pack territory of 6,4 km^2. The openness of the territories and their comparatively small dimensions lead frequently to aggressive interaction between neighbouring packs, usually accompanied by very wolfish group howling and ending always with the larger group chasing away the smaller. In areas of low prey density, wolf packs usually consist of pairs or, at most, groups of three, and the territories are much larger, on average 13,4 km^2.

Also characteristic of dogs is the inclination towards monogamy. Within each sex in a wolf pack, a hierarchy is generally well established, with frequent displays of dominance and submission. The dominant female discourages all but the dominant male in the pack from mating with her by either moving away or snarling when approached.

Because the habitat is so limited, particularly at present, there are few opportunities for dispersing Ethiopian wolves to start a new pack in an unoccupied territory. As a result, young males tend to stay with their natal packs.

At sunrise, on a frozen clump of grass in its harsh mountain habitat, an Ethiopian wolf snatches the last few moments of sleep before waking to its daytime search for food.

Females are more likely to disperse, but they are usually forced to eke out an existence in narrow ranges between territories, hoping for a vacancy in a neighbouring pack. Such openings are few and far between as the dominant female, in the event of her death, is likely to be replaced from within the pack by her dominant daughter.

This raises another point. Surely, if the daughter of the deceased alpha female becomes dominant, she will be likely to mate with her father, creating the danger of inbreeding. In reality, alpha females are not faithful to their mates and are so receptive to wolves from other packs that – in a study in Bale Mountains National Park – it was found that 70 per cent of the matings witnessed involved outside males. Mating outside the monogamous bond is far more common than is generally realised. (It will be discussed again in connection with the aardwolf.)

Unlike felids and hyaenids (except for the aardwolf), canids are seasonal breeders. Another canid characteristic is the copulatory tie, where male and female may be joined for 15 minutes. The Ethiopian wolf mates between August and November. The pups are born between October and December, but only about 60 per cent of the packs produce pups in any year. From two to six pups are born in a hole in the ground or small cave. Lower-ranking females in a pack

Scent-marking is achieved in true dog style (above right), by raised-leg urination. The dominant female in the pack usually leads the pack's patrols and its territorial marking.

Dominance (opposite) is well-established in packs. Not all packs produce pups every year, and the privilege of producing a litter is reserved for the dominant female (above left).

sometimes show signs of pregnancy although they do not give birth or may desert the pups at birth. In about half these cases, the females then join the dominant female and help to nurse her pups.

At about five weeks, the pups' milk diet is supplemented with meat provided by pack members and, at about ten weeks, the pups are weaned. Up to the age of six months, they are totally dependent on adults for food, and adults continue to feed them until they are a year old. Full sexual maturity is reached during the second year.

MONOGAMOUS AND LESS MONOGAMOUS

Jackals and aardwolf

THE JACKALS

Of all the so-called monogamous animals described in this book, the jackal appears to conform best to the model. Typically, jackals pair for life or, at the very least, for several years. If one of the pair dies, it is likely that the territory will be taken over by another pair before the widowed animal can find another mate. The small amount of sexual dimorphism in the family is a characteristic of monogamous animals.

The jackal couple jointly defends its territory and marks it with urine, faeces and by vocalization, hunting together and sharing the responsibility and work (sometimes with helpers) of raising their young. Intruders are chased away by the same-sex resident; the partner of the opposite sex does not get involved. The male will feed the female while she is confined to the den nursing their pups. Offspring from the previous year's litter may help the parents significantly by regurgitating food for the young and by guarding the den. In the woodlands of the Serengeti, non-parent helpers are found in most groups of black-backed jackals.

Black-backed jackals whelp from July to September, which is the time when grass rats, their main source of food, are most abundant. The pups emerge from the den when they are 3 weeks old and are weaned when they are as

Black-backed jackal pups (top) enjoy the benefits of a two-parent household. The male jackal feeds the nursing female while she is confined to the den and shares the responsibility of feeding the young. Regurgitated meat for the young jackals (right) is supplied not only by the mother jackal but by siblings from the previous year's litter.

young as 8 or 9 weeks of age. By 14 weeks, they are ready to leave the den and move around with the adults. The first 14 weeks of their lives are the most important in the survival of these pups, and it is within this period that most deaths occur, either from starvation or predation. The presence of non-breeding helpers makes a significant contribution to the number of pups that survive; each helper has been found to increase the average number of pups that survive by 1,5 pups per litter. The helpers are closely related to the pups. Indeed, in most cases they have the same parents – another example of kin selection.

Close by on the open plains of the Serengeti, golden jackal pairs also employ helpers. However, the greater amount of food this brings to the dens and the smaller

Golden jackal pups in the Serengeti are born later in the rainy season than black-backed jackal pups and face death not through starvation but from exposure, illness and flooding.

amount of time that the den is unguarded do not seem to improve the rate of survival of the pups. Golden jackals whelp later in the year (December and January) than do black-backed jackals, this being the time when they can depend on wildebeest afterbirth and an abundance of gazelle fawns to feed their pups. In the rainy season, the pups are vulnerable not to starvation but to exposure, illness and drowning from repeated flooding of the dens.

Canid litters are generally larger than those of other carnivores. A jackal litter usually numbers from four to six pups, and a determining factor in its size may be the presence of yearlings as helpers, passing on their genes through kin selection and learning about parenting before going off and themselves becoming parents.

Whether or not it is better for a yearling to leave and find its own territory, or to delay its departure from home ground, depends on several factors including the availability of food and variation in population density from year to year. For example, at low population density and high food availability it might be better to emigrate early, whereas in a saturated population it is better to stay home and gain experience by helping to rear the young.

Black-backed jackals, more than others, will leave their territories in groups to scavenge from large carcasses. In southern Africa, as many as 50 black-backed jackals have been reported around a large carcass. These are not members of one large social group, but a temporary concentration of several groups feeding on a particularly large food patch. Residents will try to keep intruders away, but a large carcass may attract jackals from several territories, making it impossible for the residents to maintain exclusive use of the food.

Golden jackals have not been seen in these temporary aggregations, although some stable groups comprising more than 20 animals with several breeding females have been recorded in Israel. To the best of my knowledge, there are no records of large concentrations of side-striped jackals.

The side-striped jackal is the least well studied of the jackals, but its basic social system appears to be the same as that of the other two species. There is still so much to learn and understand about these adaptable and intelligent animals.

Resident jackals are not always able to keep intruders away, epecially from large carcasses. These confrontations may look violent but they are normally ritualistic, and serious injury is uncommon.

AARDWOLF

An aardwolf pair and its latest offspring occupy a territory of 1–4 km², depending on the density of termites. Territories are marked, hyaena-style, by pasting. Although males mark more frequently than females, the average for both sexes is about 2 pastings per 100 m travelled, which translates into some 200 pastings deposited in a territory each night.

As with the brown hyaena, the smell of the resident aardwolves saturates the territory. If a resident encounters an intruder, particularly one of the same sex, it raises the long mane along its back and chases the intruder to the border. Intruders usually escape. Fights are rare, except between males in the mating season, when dominance becomes important. The aardwolf's cheek teeth have been reduced to a few small, simple pegs; it does not need teeth to catch and eat termites. However, its canine teeth are well developed and can inflict serious, even fatal, injury in a fight. These fights are accompanied by deep roars, quite out of keeping with the aardwolf's delicate appearance.

Aardwolf young do not stay on to help feed the new cubs. At about the time the next litter is born, they take off.

Like the jackal, and unlike the other members of the hyaena family, the aardwolf is a seasonal breeder. In southern Africa, mating between aardwolves takes place in early July and, after a gestation period of 90 days, one to four cubs are born in a den. They emerge from the den at about four weeks, begin foraging for termites around the den at nine weeks, and are weaned at four months, when the denning period ends and the cubs become independent.

The male aardwolf helps to raise the young by guarding the den against predators, particularly jackals, probably their greatest enemy. Males may spend up to six hours a night guarding the cubs while the mother is out foraging. Dens with male guards are three times as secure as dens where the male is absent. Superficially, the aardwolf appears to conform to the common concept of domestic bliss – but nature, through the rule of survival of the fittest, prevails. Males maximize their reproductive success by being as promiscuous as opportunity allows. Indeed, the monogamous aardwolf is highly unfaithful and, among mammals, is next only to humans when it comes to being made a cuckold!

At the start of the mating season, male aardwolves make scouting trips into neighbouring territories. At first, they do not paste during these excursions, but as the season progresses and the females come into oestrus, the more aggressive and dominant males start to paste in their neighbours' territories, sometimes pasting even more frequently inside a neighbour's territory than inside their own. This is possibly intended to intimidate the resident male and to impress the resident female. Dominant males frequently mate with the females of less dominant males. They may even displace a resident male when he is mating with a female.

The females, meanwhile, play a dangerous game. Since the male has no reason to guard another male's cubs, the crafty female mates with both males, making the paternity of her cubs uncertain. This improves the genetic diversity of the cubs and keeps the cuckolded male in attendance. He is more likely to stay around and help the female raise the cubs when there is a good chance that the cubs are his.

Marking their relatively small territories is a major occupation of aardwolf pairs, and the pastings of intruders are carefully smelled (right). Given the patterns of aardwolf social life, the paternity of pups (opposite) is often uncertain.

HIGHLY SOCIAL

Lion, spotted hyaena and wild dog

Within each of the three families of carnivores described in this book – Felidae, Hyaenidae and Canidae – one species is highly social. However, the social system and the ecological forces that have fashioned that system are different for each species. The obvious explanation for the social nature of these animals is that it improves their hunting efficiency. This is true to a point, but it is certainly not the driving force behind their social organization.

LION

Male lions in prides are often depicted as the prototype of the male chauvinist, literally the 'king of beasts'. The females kill the prey and the males eat first. The males have prime position in the shade, mate whenever the mood takes them and play no role in raising the young. They proclaim their dominion by roaring.

This is a simplified picture of the true state of affairs. The core of a lion pride is a group of 2–12 closely-related adult females (mothers and daughters, sisters and aunts) that occupy a territory. These females are not together all the time; individuals tend to go off on their own, or, more often, with one or two other females. All members of the pride have friendly meetings with each other from time to time. The females of a pride co-operate in several ways, most obviously when hunting (although the amount of co-operation shown in hunting is usually quite limited) and in defending the territory.

A large pride (right) drinks together. The core of a pride is a closely related group of 2–12 lionesses.

In joyless copulation (approximately once every 25 minutes over a period of 4 days), a female exhibits the antagonism often seen during the early mating period. Once a female comes on heat, the first male lion that finds her becomes her consort and tries to keep other males away.

Other, more subtle and more important, forms of co-operation are: synchronization of the birth of cubs, communal rearing, including suckling, of cubs and the joint defence of cubs against strange males. Females in a pride pool the cubs from four to six weeks of age to form a stable crèche until the cubs are a year and a half to two years old.

All male cubs leave the natal pride at two to four years of age. In the Serengeti, they become nomadic; in Kruger they are 'floaters', very often close to the territory they grew up in. They forge close relationships with other young males, usually, but not always, the ones they have grown up with and are closely related to. With no pride or territorial commitments to attend to, these sub-adult males concentrate on feeding and growing big and strong. They form coalitions that may be larger than the two- or three-member cheetah

coalitions; on occasion, as large as seven members. On reaching full maturity, they are ready to take over a territory.

The territory may coincide with that of a pride of females, or it may be larger, encompassing the territories of several prides. If there are other adult males in the territory, the new males must displace them, often violently. Once the new coalition of males has taken over a territory and gained tenure, they perform the dastardly act of infanticide. Being much smaller than the males, the females are usually unable to prevent the massacre of their cubs, but they do try to gang up against the males and are sometimes able to protect their cubs, particularly if the cubs are over one year old.

If they lose their cubs, the females come into oestrus within a few days and mate with the invaders. Contrary to popular belief, there is no dominant male lion in a pride.

Three small cubs are suckled (above) by a female that might not be their mother. The females of a pride tend to give birth in synchrony, and combine to raise their cubs and defend them against infanticidal males.

On the prowl (opposite) a territorial male lion spray-marks a bush to proclaim ownership of his territory. After a territorial takeover by a coalition of two males, a lion cub lies dead in the Kalahari sand (right), victim of the infanticide that is perpetrated by the conquerors in order to sire their own cubs by forcing the females into oestrus again.

The members of a coalition are of equal status, as are the adult females. When a female in a pride comes on heat, the first male that finds her becomes her consort. He sticks close to her, not letting her get more than a metre away from him, and tries to prevent other males in the coalition from approaching too close to her. However, there is no overt fighting between the males. As they are usually closely related, those males that do not mate successfully with the females, share some of their genes with the cubs through kin selection.

Even if they are not related, it is important for the maintenance of the territory that the males maintain a strong coalition. There is no point in attacking and injuring a potential ally. Better to wait and hope to get in first with the next female. As the oestrus cycles of the females are usually synchronized, there is a good chance of this happening. Mating extends for about 4 days, during which time copulation occurs roughly once every 25 minutes. If the female fails to conceive, she will come into oestrus again in 16 days. If she does conceive, she will produce from one to four cubs after a three-and-a-half-month gestation period. Long-term studies of Serengeti lions have shown that, compared with single male lions and pairs, coalitions of three or more male lions can easily gain tenure of prides of females, retain tenure for longer and mate with several different females, and that their offspring have a higher rate of survival.

When the females give birth again, the new males protect the cubs. If it were not for the infanticide committed earlier, the new males might have to wait two years before they could sire their own cubs. As the period of tenure of a pride by males is not usually more than three or four years, they might not be able to raise any of their own offspring. Although females mate with new males immediately after a takeover, they may not conceive for several months – a testing period for the new males in which they can prove themselves worthy of holding a territory for an extended period. If they should fairly soon be ousted by a coalition of stronger males, the females will not have wasted time and energy in producing offspring that will be killed. Time is on the side of the females, who may continue to produce cubs for up to 11 years.

Male lions are not the lazy chauvinists they are reputed to be. In woodland savannahs particularly (as described earlier), male lions are active hunters. Moreover, the task of maintaining the territory demands regular patrolling. During these patrols, the males announce their presence by roaring and scent-marking bushes with sprays of urine. Should intruders be encountered, the male lions may lay their lives on the line to defend their home turf in an impressive display of co-operation and teamwork.

Aside from any question of hunting advantage, an important reason why females form social groups is to reduce the chances of their cubs being killed by nomadic males in pride take-overs, particularly when the cubs are larger and more mobile. The Serengeti studies referred to have shown that a lioness has a better chance of raising her cubs in a pride of three or more females than in a pride of one or two females. Mothers keep their cubs in a creche and form maternity groups that are effective in defending the cubs against infanticidal male lions. These females are close relatives – another example of kin selection in practice.

In coalitions (opposite), usually of 2 to 4 and, sometimes, as many as 7 members, most of them related, male lions co-operate to secure a territory and gain mating rights – or else to enable those with similar genes to mate. Once secured, a territory must be regularly patrolled (above) and ownership proclaimed by roaring.

Furthermore, lionesses are better able to defend their hunting territories against other female groups if they do so in a pride. Large prides dominate smaller ones, and females attack and sometimes kill their neighbours. However, not all the lionesses in a pride participate equally in the defence of the territory. Some individuals consistently lead the attack, exposing themselves to greater risk, yet all benefit equally if the intruders are successfully repelled. This has nothing to do with the age and experience of the animals and seems to be based on individual characteristics. The lead females appear to recognize that some of their members are laggards, but do not attempt to punish them. It is difficult to explain why this is so and how such behaviour could evolve.

SPOTTED HYAENA

Although the male lion is not really a 'beastly' king, the female spotted hyaena may well be 'the Queen'. Among spotted hyaenas, the females are larger, dominant and more aggressive than males. Spotted hyaenas live in female-dominated groups, called clans, that defend group territories. All young males leave their natal or birth clan at about two years of age and go off to find their fortune in another clan. The young females are recruited into their natal clan, where they form part of a linear hierarchy in which the daughter inherits the status of the mother.

Immigrant males are also arranged in a hierarchy. New immigrants must join the clan at the bottom of the hierarchy, and are dominated not only by the adult females, but also by their cubs. They slowly work their way up the social ladder, spending much time attempting to develop amicable relations with the adult females. Over several years they may succeed in mating with some of them. However, no matter how long a male stays in the clan and how successful he is at securing matings, the highest-ranking male has a lower status than the lowest-ranking female.

As well as being larger and more aggressive than the male, the female spotted hyaena has high levels of the male hormone testosterone. This is the case even in female foetuses. In the development of the foetus of every mammal, the histology of the reproductive organs is formed at an early stage, the direction in which the reproductive organs develop being determined by the hormones secreted. The excessive amounts of testosterone in female spotted hyaenas is

responsible for their strange male-like sex organs, the clitoris enlarged like a penis and the labia of the vagina fused to form pseudo-testes. The birth canal is enclosed in the clitoris so that the female gives birth through this appendage. Contrary to popular belief, the spotted hyaena cannot change sex and is not hermaphrodite.

Because of its unusual anatomy, the birth canal in the spotted hyaena is twice as long as that of mammals of similar size. Still births and abnormal labour, which may result in death, are relatively common among young females giving birth for the first time. The umbilical cord is much shorter than the birth canal and so becomes detached from the placenta as the foetus moves down the birth canal, cutting off the oxygen supply to the foetus. As the width of the opening of the clitoris (2,2 cm) is much smaller than the diameter of the foetus's head (6–7 cm), it takes some time for the foetus to be forced through

and it may die from lack of oxygen before it can emerge. The opening of the clitoris is torn during the first birth making subsequent births much easier.

Mimicry of the male's reproductive organs by the female has been incorporated into an elaborate meeting ceremony. Two spotted hyaenas from the same clan stand head to tail, each raising the inner hind leg and sniffing at the other's erect reproductive organs. In essence, each is exposing its most vulnerable parts to the other's lethal teeth. Only two animals that know and trust each other will dare do this; males hardly ever have the courage to expose the 'Full Monty' to adult females! This expression of trust and solidarity helps reinforce the close bonds that exist between members of a clan that often co-operate in dangerous activities such as attacking gemsbok and defending the territory against intruders, and when mobbing lions (their major competitors for food).

Clan size and the size of territory vary markedly according to the amount of food available. In the arid and poorly productive Kalahari, 5–10 adult spotted hyaenas may inhabit a territory of over 1 000 km^2, whereas the phenomenally productive Ngorongoro crater supports clans of up to 80 individuals in territories as small as 40 km^2. Fights over territory are common, and fierce and fatal battles take place. In the Serengeti, with its migratory wildebeest and zebra populations, large clans defend small territories, but when prey is absent from the territory the clans have to commute to the migratory herds. A territory, thus, may be filled with non-residents from prey-poor territories when the migratory herds are there. Non-residents are submissive to and keep away from residents, and at kills they wait at a distance until the residents leave, scavenging in classic lion-hyaena style.

Head to tail, two females engage in the elaborate meeting ceremony in which each sniffs the other's male-like sexual organ. The pink scar on the clitoris of one of these females shows that she has given birth at least once.

Like lions, members of a clan separate from each other. A spotted hyaena may be on its own one night, with three members of the clan the next night and with ten others the night after that. Within a clan, the females form coalitions around their mothers. These may eventually become so large that fission takes place. The less dominant coalition breaks away and tries to set up its own clan elsewhere.

The focus of activity in a spotted hyaena clan is the communal den. The females in a clan keep their cubs together, but unlike brown hyaenas and lions, do not suckle each other's cubs. For the first 9 months or so, the cubs stay at the den and their major source of food is their mother's milk. They are weaned only at 14–18 months.

Spotted hyaena females produce the richest milk of any terrestrial carnivore. Their dominance ensures that they take precedence over males at carcasses; they eat quickly and get back to the den to nurse the cubs. Amongst the females, dominant animals take precedence over submissive ones and have a higher rate of reproductive success than their subordinates: they give birth to cubs at shorter intervals and their cubs have a better chance of survival.

Females give birth to only one or two cubs, very occasionally three, after a gestation of about 16 weeks. Most unusual for carnivores, the cubs are born with their eyes open and canine teeth fully erupted. Within minutes of birth, they often engage in a protracted and serious battle that establishes which cub is dominant and may even lead to the death of the weaker cub. This is particularly so if both are female. The dominant cub is able, largely, to control access to the mother's milk, and a cub that manages to kill its sibling will greatly improve its own chances of growing and reaching independence.

Like the other hyaenas, the spotted hyaena marks its territory by scent marking through defecating and pasting. However, the spotted hyaena does not produce the elaborate pastings of the brown hyaena, nor does it paste as often as the other hyaenas do.

What it lacks in chemical communication skills, the spotted hyaena makes up in vocalizing. It has, probably, the most complex vocal system of all carnivores. There are 14 distinct spotted hyaena vocalizations that have been identified, of which the whoop call is the one most often heard. It is one of the sounds characteristic of the African night, and is audible over several kilometres.

Spotted hyaenas can recognize each other's whoops and use them as a rallying call to scattered clan members to help defend the territory, form a hunting party, or to attack competitors, particularly lions. The whoop is also used as an individual display. Dominant immigrant males whoop often and elaborately as a way of making their presence known in the territory. The famous giggle or laugh is a sign of submission, often made when several hyaenas are feeding on a carcass.

In a strange mode of behaviour known as 'female baiting', four (usually submissive) male hyaenas attack a female.

WILD DOG

The wild dog is the most social of the large African carnivores.

Unlike lion and spotted hyaena social groups, where the members tend to come together and then go off in sub-groups for a while, a wild dog pack stays together. The unplanned separation of one of the members from the pack for even a short time is stressful, and the lost animal will attempt to reveal its location to the others by making a repeated hoo call, which sounds more like a bird than a dog.

A pack may be as small as a pair or number as many as 50 dogs. A typical pack, however, is made up of 4–8 adults, 2–6 yearlings and 5–11 pups. Although in the population as a whole the sex ratio is about 50/50, packs may be pre-dominantly male or female. Packs are formed when single-sex groups, normally brothers or sisters, break away from their natal pack and join up with a similar group of the opposite sex or locate a pack that has lost adults of one sex. Sometimes, a group of males may entice a single female from her pack, or packs may split when they are too large. Typically, same-sex adults in a pack are related to each other, but not to the adults of the opposite sex. Break-away groups generally settle close to their natal packs, but sometimes dis-perse over hundreds of kilometres.

For most of the year, a pack ranges over a far larger area than would be expected of animals of their size. In woodland savannas in the Kruger National Park in South Africa, Hwange National Park in Zimbabwe, Moremi Game Reserve in Botswana and Selous Game Reserve in Tanzania, the home range of a pack is normally about 500 km^2, and may be as large as 930 km^2. On the plains of the Serengeti, with its migratory herds, a home range may be as large as 1 500–2 000 km^2.

Members of the wild dog pack stick together – unlike lions and spotted hyaenas, which tend to come together and part from time to time. Pack size varies greatly, but averages around 12 wild dogs.

A beta female's small pup (above), killed by an alpha female.

At the den (below) an adult dog regurgitates meat for the hungry litter. There is great competition for food at feeding-time, and no room for the weak or timid.

However, home ranges shrink to about 10 per cent of the range during the three- to four-month annual denning period. Home ranges of neighbouring packs may overlap considerably, but the packs rarely meet up. When they do, they are aggressive towards each other, the larger pack dominating the smaller one. Fatalities from such encounters have been recorded.

True to the canid model, the wild dog is a seasonal breeder and the pups are born when food is easiest to procure. In southern Africa, the pups are born in late May to early June, when impala are probably at their most vulnerable. This coincides with the end of the impala rut and the onset of the dry season, when not only are many of the rams in poor condition, but the ewes and lambs are also beginning to feel the pinch.

On the Serengeti plains, wild dog pups are born in the rainy season when thousands of wildebeest calves are available. In each pack, only the top dog in each hierarchy (the alpha male and the alpha female) usually breeds. Consequently, wild dog females generally

produce large litters, on average 10–12 pups. One alpha female in Kruger produced litters of 21, 18 and 13 pups in three years, a total of 52 pups. The other members of the pack help to raise the alpha pair's pups. Providing milk for so many pups is a daunting task and, not surprisingly, the lactation period is quite short. For the first three weeks of their lives, the pups rely entirely on their mother's milk. While she is confined to the den, the alpha female is fed by other members of the pack.

The alpha female gives birth to large litters. For the first 3 weeks of their lives, she tends to them in the den, but after that nurses them above ground. Pups are weaned at 8–10 weeks and the denning period lasts about 12 weeks.

During the first three weeks, the mother spends most of her time inside the den with the pups, emerging only to be fed when the pack returns from the hunt. From about three weeks of age, the pups begin to emerge from the den and are fed regurgitated meat by other pack members to supplement their milk diet. By eight to ten weeks, the pups are weaned. Even before this, the mother often goes off hunting with the pack, but one of the other adults stays behind to guard the pups. The assistance given by the other pack members is vital for the survival of the pups and, other things being equal, the more helpers there are in the pack, the more the pups that survive.

Sometimes, a second, or even a third, female in the pack breeds. Recent studies of wild dogs have shown that this is not as uncommon as was thought. In Kruger, in nearly half of the packs containing more than one adult female, a second female was found to breed. In all cases, the subordinate female had a smaller litter than the alpha female, and her pups were born after the alpha female's. The fate of the pups seems to depend largely on the attitude of the alpha female. This varies from outright aggression by the alpha female towards the subordinate female and her pups (to the point of killing the pups), to such total acceptance of the pups and their mother that the females share the same den and even suckle each other's pups.

It is not understood why alpha females behave so differently towards subordinate females. When the alpha female is aggressive, the other pack members are far less likely to feed the subordinate female, forcing her to go off hunting with the pack, leaving her pups unprotected – from the alpha female as well as other predators. Not surprisingly, less than 10 per cent of pups that reach one year of age are the offspring of subordinate females.

With ears pricked up, two wild dog pups watch for the return of the adults from the hunt. The pups are fed, guarded and reared by the whole pack. At times of respite, particularly in summer, wild dogs may rest up for the day near a waterhole (opposite), cooling themselves now and again by bathing.

It should also come as no surprise that subordinate males will sneak a mating if they can. Genetic studies of Kruger dogs have shown that 10 per cent of pups were not offspring of the alpha male. In one case, a litter was sired by two males.

The attainment of alpha status is obviously important for a wild dog. During the mating season, the harmony of the pack is disturbed, and aggression between animals of the same sex, particularly males, is common. At such times, members of the pack may inflict nasty injuries on individuals. The hierarchy, once established, is usually long lasting, and the alpha female, in particular, is very unlikely to be usurped. Recent studies have shown that, as with humans, the high-flying dominants pay for the privilege by having higher levels of stress hormones.

Wild dog packs fluctuate in size far more than lion prides or hyaena clans. One cause of this is the high mortality rate of pups in the first year, when about two-thirds usually die. In addition, the mortality rate among adult dogs is about 35 per cent per year. A third reason is the high rate of movement to and from packs. A pack may enjoy a few years of prosperity and average over 20 dogs, only to become extinct within a year or two. The size of wild dog groups, unlike that of lions and hyaenas, is determined by causes other than the availability of food.

In Kruger, 80 per cent of the wild dog population is under four years of age and the oldest dog known is nine years of age. The wild dog's extraordinarily high daily energy consumption rate (described in the previous chapter), combined with its vulnerability to other predators – part of the subject of the next chapter – add to its 'Endangered' status and make it difficult to manage effectively.

COMPETITION BETWEEN SPECIES

the struggle for supremacy

Apart from the scavenging of hyaenas and jackals, it was long held that there is little interaction between different species of predator and that the distribution and numbers of each species are regulated purely by food supply. Each species in what scientists call the 'large carnivore guild' goes about its own business, or so it was supposed, without worrying too much about what the others in the guild are up to.

Although the size of lion and spotted hyaena populations generally corresponds closely with the amount of prey available, the same is not always true of other carnivores. For example, there are only 250–450 wild dogs in the relatively bountiful area of the Kruger National Park whereas the numbers of spotted hyaenas and lions are in the region of 2 500 and 2 000, respectively. As described earlier, there is far more overlap between the niches of carnivores than might appear at first glance. Lions and spotted hyaenas (at one level), leopards, cheetahs and wild dogs (at another) and the three species of jackal and the caracal (at a third) have similar diets. The various species are often very interested in one another's activities. They often interact and even have an effect on the numbers of each species.

In this chapter, the intricate and important relationships between some of the large African predators are described, along with the significance of such relationships in helping us to understand the size of the population and the distribution of certain predator species.

A caracal may take its kill into a tree (opposite) to escape competition from hyaenas and jackals. Interactions between predators may have a significant effect on the behaviour, numbers and distribution of the various species. At the scene of a kill, black-jacked jackal and spotted hyaena (top left) scrap over the spoils. Elsewhere, a brown hyaena and black-backed jackals feed together (middle left). A leopard is seen as enough of a threat for wild dogs to expend valuable energy treeing it (bottom left).

LION VERSUS SPOTTED HYAENA

The battle of the giants

Lions and spotted hyaenas are the top predators in the African carnivore guild, and usually feed off the most numerous large herbivores in a particular area: wildebeest and zebra in the Serengeti and the Ngorongoro Crater, gemsbok and wildebeest in the Kalahari, buffalo, zebra, wildebeest and impala in Kruger. They might select somewhat differently from the prey populations; for example, more adult gemsbok are killed in the Kalahari by lions than by spotted hyaenas, which restrict themselves mainly to catching and killing gemsbok calves. However, the similar food habits of lions and spotted hyaenas have given rise to a fierce rivalry that, at times, can spill over into outright attack. By virtue of their superior size, lions are the dominant party, but no one who has seen the spotted hyaena squaring up to its large and powerful rival would call it a coward.

One night in the Kalahari, we were following an old spotted female hyaena when we came upon a lioness feeding on a freshly-killed gemsbok. After pacing up and down for a few minutes at a healthy distance from the lioness, the hyaena started to whoop. Normally, a hyaena whoops about 5 times in a bout. This old female whooped 17 times. Almost immediately, a hyaena replied, and, within two minutes, four other hyaenas from her clan appeared. After noisily greeting each other, the hyaenas closed ranks and approached the lioness, three advancing shoulder to shoulder, the fourth one, a youngster, bringing up the rear. Until the reinforcements appeared, the lioness had paid scant attention to the old female hyaena. Now, however, she dragged the carcass deeper under the tree before continuing to feed.

The hyaenas, emitting the most intimidating series of lows, hoot-laughs and whoops, their ears cocked, their black tails bristling and curled over their backs, kept on approaching. The lioness sat up and faced them, trying to intimidate the hyaenas by growling fiercely, but her posture – ears flattened and teeth bared – was far more defensive than the confident attacking postures of the hyaenas. When the hyaenas were a metre from her, the lioness lunged at them, giving vent to a loud roar-growl in a desperate attempt to retain her hard-won prize. After only a flicker of hesitation, the hyaenas closed in. The lioness lost her nerve. She spun around and, crashing through the branches of the tree, beat a hasty retreat, leaving the little-eaten carcass to the hyaenas.

For spotted hyaenas, it is not always so easy. More often than not, lions are able to keep a carcass from hyaenas, relinquishing it only when they have had enough. Sometimes, the roles are reversed and lions steal hyaena kills. In the

Finding safety in numbers, 13 spotted hyaenas mob a lioness (above left) and are even able to cause a large male lion (above) to move on.

Ngorongoro crater, where hyaenas greatly outnumber lions, some male lions may live exclusively for a while on their thefts of hyaena kills. A large adult male lion may put 20 hyaenas to flight.

The well-known hyaena researcher Hans Kruuk often watched spotted hyaenas kill a wildebeest at night only to be dislodged by lions. The hyaenas would then wait around for the lions to finish. At sunrise, the tourists would come on the scene and remark on how the scavenging hyaenas were waiting for the lions to leave their kill! Although hyaenas are very respectful of male lions, with lionesses and young lions it is more of a numbers game. If the ratio of hyaenas to

lionesses is around four to one, the hyaenas usually have the upper hand. If it is less than that, the cats do. This may be another reason why lionesses are social animals; it helps them to safeguard their kills from spotted hyaena.

More surprising than the conflicts over kills between lions and spotted hyaenas are those that take place away from food. In such instances, particularly in the Kalahari, spotted hyaenas seem to go out of their way to engage with lions. They may spend as much as an an hour harassing them. On three occasions, I have seen lions jump into trees in order to escape from spotted hyaenas.

Lion mobbing is obviously a dangerous occupation for spotted hyaenas. The stakes are high and a false move or miscalculation in the presence of such formidable foes can lead to death. In areas like the Ngorongoro Crater, where prey in the form of wildebeest and zebra are abundant, hyaenas are far less prepared to take on lions than they are in the prey-sparse Kalahari. Hyaenas that lose a kill to lions in Ngorongoro can easily make another kill. Kalahari hyaenas that lose a kill to lions might have to travel over 50 km to make another one. It is worth it for spotted hyaenas in the arid environment of the Kalahari to make life as

developed in the Crater and lions were their preferred host. As a result, the lions developed devastating skin infections and were unable to hunt. At that time, there were 385 spotted hyaenas in the area. If ever there was an opportunity for spotted hyaena to squash their great rivals, this was it. In the event, the lion population recovered and, within a decade, had returned to its original size.

The spotted hyaena population may not show the same resilience. In the 1970s, there was intensive culling in the central district of the Kruger National Park of both lions and spotted hyaenas. Over five years, 429 lions and 364 hyaenas were culled in a misguided attempt to reverse the negative population trend in wildebeest and zebra. The lions were back to their previous numbers within a year, but the spotted hyaena population recovered far more slowly. The spotted hyaena population had not regained its size even a decade later.

The varying ratio of lions to spotted hyaenas in different areas is intriguing. In the Kalahari, for example, lions outnumber spotted hyaenas by about 2 to 1; in Kruger, their numbers are about equal; and, in the Serengeti, spotted hyaenas hold the numerical advantage. The distribution and type of prey is probably the main reason for this. In the Kalahari, prey are thinly distributed and both species live at low densities. However, lions are far more adept at killing adult gemsbok, the most common and widespread prey in the area. Spotted hyaenas have to find the calves, which are much fewer. In Kruger, the prey are more sedentary and evenly distributed, so that neither species has an advantage.

In the Serengeti, the spotted hyaena is adapted to following the migratory wildebeest and zebra, an abundant food source. When the migrants leave, the less mobile lions have to make do with the relatively small numbers of sedentary prey such as topi and hartebeest.

difficult as possible for lions that come into their territories. This might encourage the lions to move on and so decrease the chances of the hyaenas losing a hard-won meal.

Although sometimes stated, it has not been proven scientifically that, in areas where they outnumber lions, hyaenas keep down the numbers of lions by preying on lion cubs. An example of this certainly *not* happening is the decimation of lions in Ngorongoro, where the number of lions dropped from 75 to 10 in the early 1960s. After exceptionally heavy rains, large swarms of biting flies

LION AND SPOTTED HYAENA VERSUS CHEETAH AND WILD DOG

An unequal struggle

Lions and spotted hyaenas may not affect each other's numbers adversely, but they can have a great effect on cheetah and wild dog numbers. In the Serengeti, a staggering 95 per cent of cheetah cubs die before they reach maturity. Spotted hyaenas and, especially, lions are responsible for over 70 per cent of cheetah cub deaths.

Most cubs are killed during their first two months of life while they are in a lair (usually in a marsh, a clump of tall vegetation, or a rocky outcrop) and during the first two weeks of leaving the lair with their mothers. Lions sometimes find a lair by sighting the mother cheetah and moving over to investigate. More often, they simply come across it while the

mother is out hunting. If the mother is present, she will try to defend her cubs by rushing at the lion and threatening it, but this is usually futile. The lions rarely eat the cubs that they kill. As the cubs grow, they become less vulnerable and are able to scatter and hide should a large predator attack. In these instances, most of the cubs escape. Cheetah lairs are less vulnerable in areas where the bush is thicker, but the cheetahs in such areas have to make do with less suitable conditions for hunting.

Apart from killing cheetah cubs, lions and spotted hyaenas (the latter, particularly) are always on the look out for a free meal at the expense of a female cheetah and her cubs. This behaviour has been given the rather exotic name of kleptoparasitism. A female cheetah and her three six-month-old cubs, which my team and I followed in the Kruger National Park for two weeks continuously, finally managed to kill an impala ram on the sixth day. However, before she and her brood were able to eat anything, the kill was snatched by a spotted hyaena. Even the rather timid brown hyaena is able to chase a cheetah away from its kill, as are leopards and wild dog packs.

It is highly likely that a cheetah cub in the Serengeti will be killed by a lion (opposite) before even emerging from the lair – at two months of age. Spotted hyaenas are also formidable enemies. This spotted hyaena (left) has gained a meal at the expense of a female cheetah and her three cubs – the cheetah's first kill in six days.

Cheetahs avoid confrontation with lions and spotted hyaenas in several ways. They hunt, for instance, at the time of day when their competitors are least likely to be active. Most cheetah kills in Kruger are made between 10:00 a.m. and 4:00 p.m. – an expensive compromise in terms of water loss, particularly in the hot summer months. On the Serengeti plains, the cheetah chooses to hunt in areas of low prey density, where other large carnivores are less likely to be found, and even refrains from hunting if hyaenas are in the area. It is forced, therefore, to coexist with lions and spotted hyaenas in smaller numbers than would be predicted on the basis of food supply alone.

Wild dogs suffer in similar fashion: they lose lives to lions and meals to spotted hyaenas. I once saw a pair of sub-adult male lions kill 7 out of 10 wild dog pups in less than 15 seconds. The pups, accompanied by 13 adults, were moving along the road when the lions rushed out of the bush. The adult dogs scattered in all directions, but the small, 14-week-old pups were unable to escape. The lions just bit the pups and left them lying in the road, some of them still kicking. They made no attempt to eat them. On another occasion, a nursing mother left her den late in the morning for a drink at a nearby waterhole and was ambushed and killed by a lioness. The lioness ate only a small portion of the dog; it then dragged the carcass under a bush and left it.

Lion predation accounts for almost one-third of wild dog deaths in the Kruger, and is the most common cause of death among wild dogs there. Like cheetah cubs, the dogs are not killed for food, but simply from aggression by a larger species towards a smaller species with encroaching food habits.

A wild dog pup at the feet and in the jaws of a sub-adult male lion, one of two that had just killed seven pups. Lions have a major impact on the density and distribution of wild dog populations.

Impala make up nearly 80 per cent of wild dog kills in the Kruger National Park. Although we do not have accurate figures for the number of impala in Kruger, there are certainly over 100 000 of them, possibly twice that number, which is surely more than enough to feed 450 dogs (about the maximum number recorded at any one time in Kruger). It seems logical, moreover, that the highest density of dogs would be found in areas where impala abound. Surprisingly – as in the case of cheetah in the Serengeti – the opposite is true. There is, in fact, a strong negative correlation between impala and wild dog densities.

An analysis of the habitat preferences of impala and wild dog has shown that some habitats are favoured by both species, but others are avoided by one or the other. A similar analysis of habitat preferences of lions and wild dogs showed that lions' preferred habitats are the ones most avoided by wild dogs. The prudent avoidance by wild dogs of their greatest natural enemy explains the rather strange distribution pattern of wild dogs in relation to their major prey species. Lions are an important factor in determining the manner in which wild dogs are distributed in Kruger and in keeping their numbers low. In large conservation areas in the rest of Africa, this also holds true: the higher the lion density the lower the density of wild dogs.

Spotted hyaenas in Kruger are merely a nuisance to wild dogs. Kruger spotted hyaenas usually forage on their own and, although they may steal kills from small packs of dogs, the dogs are generally able to keep them off by force of numbers. In addition, the thick bush makes it difficult for hyaenas to locate wild dogs feeding. Should a hyaena venture close to a wild dog den, the pack will quickly see it off.

If the dogs catch up with a hyaena, they will give it a thorough going over, surrounding it and biting it in the backside – a noisy affair because of the defensive growls and yells of the hyaena.

Large, well-managed reserves in woodland habitats like South Africa's Kruger National Park, Tanzania's Selous Game Reserve and the Moremi Game Reserve in Botswana hold stable wild dog populations. However, in open habitats such as the Serengeti ecosystem, wild dogs are under far more pressure from spotted hyaenas, which are common and are present in quite large groups. Because wild dogs consume energy at such a high rate, even a small loss of food to hyaenas is serious. Wild dogs need to hunt on average for three and a half hours a day in order to obtain their required daily energy dose. They spend the rest of the time resting in the shade. A computer model using data on the amount of energy used by a wild dog while hunting and at rest suggests that if wild dogs were to lose a modest 25 per cent of their kills, they would need to hunt for 12 hours a day to compensate. This, clearly, is impossible.

The wild dog became extinct in the Serengeti about 10 years ago. Rabies, and possibly canine distemper, may have been the final nail in the coffin, but wild dog numbers had declined even before the onset of disease while hyaenas and lions had increased as a result of favourable conditions. This almost certainly increased hyaena kleptoparasitism and, possibly, wild dog mortality from lion predation.

For wild dogs, the leopard is a different story. Whenever they encounter a leopard, the dogs chase it, and the leopard makes for the nearest suitable tree. In true cat and dog fashion, the dogs gather under the tree looking up and sometimes making prodigious leaps in the air. It might appear to be 'sport', but it must be important since it consumes precious energy. Leopards and wild dogs compete closely for food, and leopards have been known to kill wild dog pups.

The tables are turned briefly as a pack of wild dogs mobs a lioness – a dangerous, and potentially deadly, activity, but one that the smaller competitor occasionally engages in.

SPOTTED HYAENA VERSUS BROWN HYAENA

Skirmishes in the desert

That most successful of carnivores, the spotted hyaena, is outnumbered in the dry southern Kalahari not only by its major competitor, the lion, but also by the smaller and far less aggressive brown hyaena. In this rather unproductive ecosystem, the solitary scavenging brown hyaena is better suited than its more social relative to eking out an existence on widely scattered and often small food items. Because of the low numbers of spotted hyaenas in the Kalahari, the two hyaena species rarely meet. When they do, it is usually more than a casual encounter.

As both species are well adapted to scavenging, they are most likely to come across each other at carcasses. For the most part, the brown hyaena, being the more widespread, enjoys the pickings without having to compete for them with spotted hyaenas. In the vicinity of spotted hyaena dens, however, spotted hyaenas tend to monopolize the carcasses. They will take over from brown hyaenas on a carcass, even in a one-on-one or two-on-two situation.

One beautiful Kalahari moonlit night, I followed two spotted hyaenas to a hartebeest carcass at which two brown hyaenas were already feeding. Predictably, the brown hyaenas withdrew as soon as the larger hyaenas arrived. I expected that this would be the last I would see of them that night. However, the two brown hyaenas retreated only a short distance. From time to time over the next four hours, one of them would come right up to the spotted hyaenas, its mane raised in typical brown hyaena fashion, and the spotted hyaenas would move towards it. It would retreat, then stop and give vent to a very deep short growl. This would stop the spotted hyaenas in their tracks and, for half a minute or so, the three animals would stand still, the brown hyaena growling intermittently, the spotted hyaenas looking around in all directions

Scavengers both, spotted hyaena and brown hyaena meet most often at a carcass (right). On rare occasions, they will feed together for a while (opposite). Usually, however, the brown hyaena is evicted.

except, it seemed, directly at the brown hyaena. Eventually, at sunrise, the two spotted hyaenas – tired out by the persistent, albeit low-keyed, protest of the brown hyaenas and, probably, having eaten their fill – departed, leaving the two brown hyaenas with the remains of their carcass.

I subsequently witnessed several similar incidents. In all cases, the spotted hyaenas were clearly dominant. Yet it seemed to me that by persistent low-keyed pestering of the spotted hyaenas, the brown hyaenas caused them to abandon their food earlier than they would otherwise have done. This is similar to the flamboyant and noisy exhibitions made by spotted hyaenas in the vicinity of lions. There is, however, an important difference. Spotted hyaenas never lose large amounts of food to brown hyaenas, whereas lions sometimes do lose out significantly to spotted hyaenas.

Away from food, the two hyaena species do occasionally cross paths. Once again, their behaviour is surprising. More often than not, the brown hyaena will come closer, generally provoking a chase. If caught up with, it faces the spotted hyaenas snarling with hair raised, ears back and mouth open. As in the encounter described above, the spotted hyaenas will stop and look off in different directions, as if oblivious. Alternatively, they will 'dance' around their captive, seemingly reluctant to attack it, but darting in from time to time to nip at it or even sniff it before stopping and looking off in different directions again. Eventually, the spotted hyaenas lose interest and the brown hyaena can slink away.

If the brown hyaena is unlucky, the encounter will escalate into a serious skirmish. The brown hyaena will be grabbed by the side of the neck and shaken vigorously, yelling in protest. In most cases, it manages to break away and escape without being pursued. If it is very unlucky (as happened in 1 of the 28 such encounters I witnessed) it will be killed. Although spotted hyaena are unquestionably dominant over brown hyaenas, which they deprive of food, harass and even occasionally kill, there is a strange attraction between the two species. Striped and spotted hyaena in the Serengeti have a similar relationship.

Driving around in the Kalahari, I saw brown hyaenas four times more frequently in areas where there were few spotted hyaenas than in the area of spotted hyaena dens. As within lions and wild dogs in Kruger, it appears that a larger and more dominant species has a negative influence on the distribution and numbers of a smaller rival. Because the southern Kalahari cannot support a large spotted hyaena population, the influence there of spotted hyaenas on brown hyaena numbers is small. However, in Kruger, which today supports about 2 500 spotted hyaenas, the brown hyaena is absent as a breeding species. The handful of brown hyaenas seen in Kruger over the last decade are vagrants from neighbouring areas, where brown hyaena are not uncommon and spotted hyaenas have been eradicated. Interestingly, Kruger spotted hyaenas eat many more small food items (normally brown hyaena fare) than their counterparts in the Kalahari.

BROWN HYAENA, BLACK-BACKED JACKAL AND CARACAL

Eating the other and the other's food

Brown hyaenas, black-backed jackals and caracals are common and widespread in the more arid regions of southern Africa. The relationships between these three species may not be intricate, but they form a web of influences and effects.

In the Kalahari, black-backed jackals are so widespread and common that they are usually the first scavengers at an abandoned carcass or the body of an animal that has died from causes other than predation. Although they are small, black-backed jackals are extremely efficient at disposing of the remains of a carcass. What they cannot eat immediately, they bury close by for recovery later. In this way, they may deprive the brown hyaena, who usually arrives later on the scene, of a substantial amount of meat. The brown hyaena gets its revenge when it steals a springhare, steenbok or other small mammal kill from a jackal. Brown hyaenas are also known to eat jackals whenever they get the chance.

Often, when a brown hyaena starts to eat a bone, a black-backed jackal appears, obviously attracted by the cracking sound. By persistent pestering, even to the extent of nipping the hyaena in the back legs, the jackal will cause the hyaena to move away a few metres with the food. Immediately, the jackal will go to the place where the hyaena was eating and pick up bone chips and the scraps that the hyaena would have eaten had it not been disturbed. Not a great loss to the hyaena, but gain enough for the jackal.

The caracal may also lose kills to the brown hyaena. I have seen brown hyaenas steal three steenbok, two springhares, two springbok and an African wild cat on different occasions from caracals, all of which provided a really good meal for the hyaena. Like the leopard, the caracal occasionally takes its kill into a tree – a most effective way of escaping kleptoparasitism – but it does not appear to have the same strength as the leopard, even in relative terms.

Black-backed jackals and caracals are also well tuned to each other's activities. Caracals eat jackals and vice versa. The caracal appears to be dominant. I once saw a caracal steal a springhare kill from three jackals. The three jackals gathered themselves and came back at the cat, but – true to its fearless nature – it stood its ground, spitting and snarling, and kept the jackals away. However, the caracal did not manage to keep its prize for too long, because a second kleptoparasite in the form of a brown hyaena came along a short while later and took the springhare away.

Although smaller than the brown hyaena, particularly in the drier western region of southern Africa, the black-backed jackal is usually the first at the scene of a kill (left). On the other hand, hyaena bone crunching can attract a jackal to the site (below), where it waits around patiently.

In many parts of southern Africa, particularly the sheep-farming districts, the black-backed jackal is regarded as a pest. Over the years, a concerted attempt has been made to eradicate the black-backed jackal in sheep-farming areas. Although this has reduced the number of jackals, it has increased the number of caracals and has created, in many ways, an even bigger problem.

In the troubled relationship between humans and carnivores, there are other and more enlightened strategies to solve problems such as jackal predation. Some of them are described in the next chapter.

CONSERVATION

the search for solutions

All the adaptations and struggles for survival discussed in this book have evolved over millions of years. Those individuals and species that have been able to adapt to changing conditions have survived and have passed on their genes to subsequent generations. In relatively recent times, however, a species has evolved that has come to dominate and change the world as no other species has done. Humans have changed the planet to such an extent in the space of the last few hundred years that thousands of species of animals and plants have not been able to adapt and have become extinct. Others are dying out at a rate more rapid than ever before.

Conservation is different things to different people. To me, it is the wise use of resources on a sustainable basis. My vision is that the carnivores of Africa be managed in an ecologically and economically sustainable manner, free from irrational and unnecessary persecution. The World Conservation Union (IUCN), the largest conservation organization in the world, has three conservation objectives: to secure biological diversity, to use the earth's resources wisely, equitably and in a sustainable manner and to guide human communities towards ways of life that are both of good quality and in harmony with nature.

What are the major issues facing the conservation of Africa's predators and what should be done to resolve them? The Species Survival Commission (SSC), a volunteer commission of the IUCN has, in the last decade, produced surveys and action plans that assess conservation status and make recommendations. It places each species in one of a number of categories, the most important of which are 'Critically Endangered', 'Endangered', 'Vulnerable' and 'Lower Risk'.

It is one thing, however, to make recommendations for action in this area. The real challenge is to implement them.

Four of Africa's most vulnerable predators: Ethiopian wolf (top left); wild dog (middle left); lion (bottom left) and cheetah (opposite). This cheetah has been trapped on a Namibian game ranch for release and radio tracking.

ETHIOPIAN WOLF

Critically Endangered

The Ethiopian wolf is Africa's most endangered carnivore and the most endangered canid in the world. It is one of several species endemic to the highlands of Ethiopia, with only about 500 individuals surviving in a few isolated populations. High-altitude subsistence agriculture and overgrazing threaten the afro-alpine habitat of the wolves. The situation is made worse by inbreeding and loss of genetic diversity, as well as by the spread of diseases such as rabies from domestic dogs.

Improved management of protected areas, particularly with regard to keeping out people and their livestock, active efforts to monitor and protect the remaining populations of wolves and the establishment of a wolf population management programme are the most important protective measures aimed at saving Africa's only wolf from extinction. A key strategy is to manage the several small isolated populations of wolves as if they are a single population, namely a meta-population, so that gene flow between the sub-populations can take place. This is a management strategy that is of great relevance to conservation programmes of the 21st Century.

Improved park patrolling by rangers, control of domestic dogs within the parks and the education of neighbouring communities are also required. All this in a very poor country where conservation is understandably (though unfortunately) low on the national agenda, infrastructure is poor and training for conservationists is practically non-existent. On the positive side, there are several Ethiopians, as well as a few expatriates sponsored by international conservation organizations, who continue to work hard and against the odds to try to save this species.

The Ethiopian Wolf Conservation Program based in the Bale Mountains National Park employs about 20 people. Field assistants monitor the wolf populations on horseback, recording their movements and pack sizes, looking for dens and counting pups. The rabies control project visits villages to vaccinate domestic dogs and to sterilize dog-wolf hybrids and unwanted dogs. Education officers produce educational materials and visit schools around the park to educate children on how best to protect their environment and the area's unique wild life. The programme assists park staff with routine management and actively supports patrols. It is looking for ways of promoting ecotourism in the area.

'The Ethiopian wolf is our country's unique heritage. Let us work together for its protection' – *translation of a road sign (opposite) in the Bale Mountains National Park.*

A domestic dog (left) belonging to subsistence farmers in Ethiopia is a close relative of the Ethiopian wolf but a potential threat. Dogs mate and hybridize with wolves and may transmit harmful diseases like rabies. To solve such problems, specialist groups of the Species Survival Commission, a volunteer arm of the IUCN, have adopted a multi-disciplinary approach.

WILD DOG

Endangered

Over the last half-century, the wild dog has disappeared from 25 of the 39 countries in which it was once recorded. Fewer than 10 populations number more than 100 animals. It is thought that there are no more than 600 to 1 000 packs in the wild, totalling only 3 000–5 500 wild dogs. Habitat fragmentation, persecution and loss of prey are major causes of the decline of the wild dog. These pressures have been experienced, however, by all large species of carnivore and none of them is in quite the same predicament as is the wild dog.

More than any other African predator, the wild dog has suffered from strongly negative attitudes. Stock farmers have felt justified in their concerted campaign to exterminate it. Until the mid-1960s, even game wardens were controlling wild dog numbers in conservation areas in order to 'protect' prey populations. Unfortunately, this attitude persists today in several areas where wild dogs and their natural prey occur.

Captive-bred dogs (above) may help to build up wild dog populations if wisely used in combination with dogs caught in the wild. The map (below) shows the difference between the past (light areas) and present (dark areas) distribution of wild dogs in Africa.

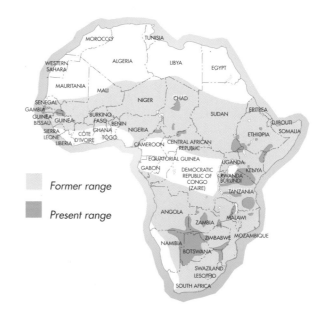

Former range

Present range

Although wild dogs are no longer persecuted in protected areas, human activity around these areas is important. A recent study in Zimbabwe's Hwange National Park found that more than half the number of dogs whose deaths had been recorded had been killed along a main road bordering the park. Similar situations are on record in protected areas in southern Tanzania and Zambia. Snaring, poisoning and shooting are also important factors in mortality rates around protected areas. Snaring is usually secondary in that the target species are antelope; for some unexplained reason, wild dogs seem to get caught in snares more often than any other of the larger African predators.

Wild dog packs roam over very large areas. Consequently, it is only in the conservation areas of about 10 000 km^2 or more, in suitable habitat, that enough packs (a minimum of nine) can be protected to form a viable population. The future of the wild dog, more than that of any other African carnivore, depends on our maintaining the integrity of the large conservation areas in which they are found. The Peace Parks initiative and efforts to create large transfrontier game parks in southern Africa is of particular relevance to wild dog conservation.

In South Africa, outside the Kruger National Park, there are a number of securely-fenced protected areas that could support one, two or three packs of dogs. It might be possible to reintroduce wild dogs into some of these reserves and manage these sub-populations of dogs as if they were a single meta-population. As wild dogs will not be able to move between these reserves, they will have to be moved artificially and in keeping with natural wild dog strategies of immigration and emigration. The basis of this is single-sex groups breaking away from packs and joining up with similar groups of the opposite sex.

Fitting a radio collar on a wild dog in the Kruger National Park (right) is part of a long-term research and monitoring programme. Once widespread throughout much of Africa south of the Sahara, the distribution of the wild dog has shrunk to a few isolated populations, mainly in eastern and southern Africa.

There are several sources of wild dogs for this meta-population in southern Africa, including dogs born and raised in captivity. There is also a limited supply of free-ranging dogs from Kruger and from unprotected areas where wild dogs are regarded as a problem. Animals caught in the wild are more suitable for re-introduction than those bred in captivity. Dogs bred and reared in captivity are often inbred, poorer hunters than dogs reared in the wild and, more importantly, are less wise to the ways of competing carnivores, particularly lions. In addition, they tend to be imprinted and dependent on people. However, dogs reared in captivity are more easily available and may be mixed with dogs caught in the wild.

In 1998, following an international workshop in South Africa on the wild dog, the Wild Dog Action Group (WAG) was established under the auspices of the Canid Specialist Group and the Endangered Wildlife Trust to improve the conservation status of the wild dog in the country. To date, as part of an attempt to establish a future meta-population, wild dogs have been reintroduced into three reserves in South Africa: Hluhluwe Umfolozi Park in KwaZulu Natal and, in the North-West Province, the Madikwe Game Reserve and Pilansberg National Park.

Wild dogs were first reintroduced into Hluhluwe Umfolozi Park in the early 1980s. Despite promising beginnings, their numbers began to drop by the early 1990s and they failed to produce young. Inbreeding is the most likely explanation for this failure as no new dogs were entering the population. New blood has recently been introduced and the results are being monitored.

In 1995, 3 male dogs from Kruger were introduced to 3 captive females from De Wildt Captive Breeding Centre in a holding camp in Madikwe Game Reserve. The 6 mixed well and produced puppies a year after their release. The pups thrived and the pack bred again, in 1997, before disaster struck in the form of a rabies outbreak. Within a short period, the pack of 24 was reduced to just 3 survivors, who were taken back into the holding camp.

This was a big setback, but an important lesson. Small populations need to be vaccinated against rabies. Research on an effective vaccination against rabies was begun and, in the meantime, more dogs were introduced into Madikwe, this time in two packs, made up from the rabies survivors, dogs from Kruger and the Northern Province and dogs bred in captivity. The urgency of the need to develop a rabies vaccination was reinforced by a second outbreak of rabies in Madikwe early in 2000.

A pack of 10 dogs made up of 3 captive-bred males from Cango Wildlife Ranch near Oudtshoorn and 2 adult females and 5 pups from Northern Province has been formed in Pilansberg, the third reserve to be included in the establishment of the meta-population. More areas are being identified, and plans are being made to build up the meta-population into a viable unit.

Even if the meta-population strategy is successful, there is only a limited number of reserves suitable for wild dogs in South Africa. The bushveld farming areas, particularly in the Northern Province, are a far larger area of potential wild dog habitat. Many stock farmers in these areas, tired of fighting the frequent droughts that plague much of southern Africa, have moved away from farming domestic animals and have introduced drought-resistant indigenous wild game species onto their farms. Live game sales, hunting and ecotourism are the three ways in which these animals generate income.

Game farmers vary in their attitude towards the wild dog. Some are tolerant, even regarding the dogs as an asset, but most still see them as a threat to their livelihood and of no value whatsoever. Although it is illegal to kill a wild dog on a game ranch in South Africa without permission from the conservation authorities – which they are extremely unlikely to give – farmers often take the law into their own hands and are seldom prosecuted for doing so. Nature conservation authorities, in an attempt to appease farmers and save the dogs, give permits for the live capture of dogs. In most cases, this merely condemns the dogs to a life in captivity. The real solution – and challenge – lies in finding ways for wild dogs and game farmers to coexist.

Although wild dogs kill animals that could be sold or hunted, they are selective hunters and weed out the weaker animals, thereby contributing to the general fitness of prey populations. Game farmers argue that wild dogs kill valuable species such as roan and sable antelope. However, of a sample of 4 127 wild dog kills from the Kruger National Park, only 18 (0.4 per cent) were sable or roan. Predation by wild dogs on these species is highly exceptional, which is why it is possible to take out insurance at a very low premium on valuable game killed by wild dogs.

The more enlightened game farmers realise that the larger the area, the more ecologically stable it is. They are taking down fences and going into partnership with neighbours by forming conservancies. This enables them to hold a wider range of species and makes it easier to accommodate carnivores, including wild dogs. Despite the ecological soundness of this method, however, it is difficult to get farmers to agree on how to manage an area and share the assets.

Although wild dogs are not trophy animals, they provide opportunities for high-quality game viewing. During the denning season, in particular, their behaviour is predictable and they are easy to find and observe. Game farmers could almost guarantee their clients a view at close quarters of pups playing and of dogs feeding pups.

As for the economics of game farming and wild dog eco-tourism, the costs of wild dog predation (approximately the cost of 1 impala a day to feed a pack of 12 dogs), are far lower than the sum of money that ecotourism could generate in a denning season of 12 weeks.

The conservation of the wild dog in game ranching areas needs to be investigated as a matter of urgency on three fronts: the biological aspects, the economic benefits and the implementation of education campaigns aimed at game farmers and other important stakeholders. Without innovative action and hard work on the part of all involved in these areas, the words of R C F Maughn might well be fulfilled and the wild dog's 'complete extermination' may ultimately be effected.

Should that happen, it will be a sad day indeed for African game and its preservation.

At home in the wild – a pack of wild dogs in Kruger, one of the last refuges of this much-persecuted animal.

LION

Vulnerable

Not surprisingly, lions are often considered a serious problem and at odds with human settlement and Africa's widespread cattle culture. Today, lions are mostly confined to the larger conservation areas, and it is on the perimeters of these areas that most conflict occurs.

It appears that there are two classes of marauding lions. A study in Namibia's Etosha National Park describes occasional raiders (members of a pride whose territory coincides with the fenced border of the park and who rarely transgress) and habitual stock killers who are non-territorial. Returning occasional killers to their territory often solves the

problem and they do not transgress again. However, most non-territorial stock-killers – usually sub-adult animals – return to stock killing and should be removed.

It may not always be necessary to kill habitual stock-killers. In South Africa, there are often quite small, fenced-off areas, many of them reserves that promote ecotourism, into which these lions can be translocated. Although such areas cannot support a self-contained, viable lion population, the value of lions for ecotourism makes them welcome inhabitants. Long-term management of lions, as of wild dogs, as a meta-population in relatively small areas is a challenge.

Although lion-hunting is unpalatable to many, it must be considered that the marauding lions have killed the cattle of peasant farmers and that they are habitual stock killers.

Why should a proportion of the considerable amount of money that some hunter is prepared to pay to shoot the lions not be channelled into communities that have suffered? As long as it is done humanely, and enough of the money is channelled into the appropriate communities and conservation areas, lion-hunting offers a solution. As conservationists, we should be concerned less with the lives of individuals than with the long-term viability of populations.

Not that this need be the final solution. Electric fencing can be very effective in keeping predators within the boundaries of reserves, and it is the responsibility of stock farmers on lands bordering conservation areas to implement sound management practices to protect their stock from predation by lions and other predators.

In the Makgadikgadi National Park, Botswana, the small and highly vulnerable lion population is in direct conflict with cattle ranchers along the Boteti River. In the dry season, large herds of wildebeest and zebra gather here because it is the only area with water. During the rainy season, the game migrates to the east, but the lions do not, and their predation on livestock increases. The local communities, who receive partial compensation from the government for livestock losses, believe that it is the government's job to protect their animals from lions and are reluctant to solve the problem in ways other than killing lions.

A proposal has now been made to implement what has been called 'reinforced demonstration', whereby local communities will be helped to build simple but adequate lion-proof enclosures into which livestock can be placed at night. Compensation (in full) will then be paid only in those cases of lion predation where it can be demonstrated that livestock was killed within the lion-proof enclosure.

In the Serengeti, a member of an unattended herd of cattle that strayed into the park falls prey to a pride of lions (left and below). Conflict between lions and cattlemen is a widespread problem along the boundaries of protected areas and needs innovative solutions, such as those that have been proposed in Botswana.

CHEETAH

Vulnerable

Like the lion, the cheetah has an IUCN status of 'Vulnerable' which is one up from 'Endangered' and means that it is unlikely to become extinct in the short term. However, it is assessed as being under considerable risk over the next two to five decades. It is almost extinct in the north of Africa and in Asia, but is still quite widely distributed in East Africa and in southern Africa.

Much has been written about the cheetah's lack of genetic diversity resulting from a history of severe population bottlenecks. It is argued that this results in poor reproduction and renders the cheetah vulnerable to infectious diseases. However, evidence for this hypothesis comes from captive-bred cheetahs and does not hold true for cheetahs in the wild.

The threat to cheetahs and other predators from larger predators, especially lions and spotted hyaenas, was described in the last chapter. It is important that this is taken into account in the management of protected areas. For example, in Kruger, the provision of boreholes undoubtedly led to an increase in numbers among species such as impala, wildebeest, zebra and buffalo, and tended also to make these species sedentary. With the improved food supply, the numbers of lions and spotted hyaenas increased. Other species, however, did not benefit. Selective grazers such as

A relaxed cheetah uses a safari car as a vantage point. Such easy coexistence is not always the case, however, and high-density tourism in game reserves may disturb cheetahs.

Getting a second chance, a 'problem cheetah' is released on a cheetah-friendly game ranch in Namibia.

roan antelope and tsessebe suffered because bulk grazers, such as zebra, encroached on their food supply. Cheetahs, wild dogs and brown hyaenas may well have suffered also, through increased competition from the larger lion and spotted hyaena populations. Kruger's policy has changed, and some of the artificial water holes are being closed in the interests of conserving species diversity.

Where large predators have been eliminated, as on Namibian and northern South African ranches and on farm and pastoral land in Kenya, as well as in parts of Somalia, the cheetah appears to thrive. There are, arguably, more cheetahs on agricultural lands in these areas than within protected areas. The success of cheetahs on ranch land is not always viewed favourably by owners of livestock and, in most areas, landowners are permitted to kill cheetahs that pose a threat to their livestock. Although justifiable in certain cases, this policy is open to abuse through indiscriminate killing. According to government statistics, 5 600 cheetahs were killed in Namibia from 1980 to 1991.

Ironically, this issue is easier to address on livestock than on game ranches (as indicated earlier in connection with the wild dog) because domestic animals are easier to manage. They can, for instance, be herded and secured in enclosures at night. A number of innovative methods aimed at curtailing cheetah predation on livestock have been developed by the Cheetah Conservation Fund in Namibia and similar organizations. One that has achieved considerable success is the use of Anatolian shepherd dogs from Turkey. While still pups, the dogs are placed with a herd of livestock and grow up with them, forming a close bond. They accompany the livestock when out grazing and protect them from cheetahs and other predators. Donkeys have also been found to be effective guards of calves of domestic cattle, even against leopards.

A second method, still in the testing stage, is conditioned taste aversion. The idea is that a predator baited by livestock that has been laced with lithium chloride will become violently ill on consuming the bait and, thus, can be conditioned to avoid the 'poisonous' prey.

The cheetah is also being made to pay for itself. ('If it pays, it stays!') To this end, the authorities in Namibia and Zimbabwe are allowing limited trophy hunting of cheetahs on private land. To be sustainable, however, a quota system such as this has to be based on sound population estimates, which are difficult to obtain for cheetahs in many areas.

LEOPARD

Lower Risk

Of all the big cats, the leopard is best able to look after itself. It is tolerant of habitat modification and able to live close to man. The collapse of the fur trade due to changing public opinion and the imposition of strict trade controls under the Convention for International Trade in Endangered Species (CITES) has, thankfully, reduced poaching.

One approach to the problem that leopards and other predators pose by killing livestock has been to translocate the offending animals to a protected area. In most cases, however, there has been no follow-up study. The animals have been left in the protected area and everyone has gone home with a warm feeling of having saved them. Unfortunately, the few follow-ups that have been done on translocations show that the animals hardly ever stay in the release area and often return to the original locality.

Whether a hunter's trophy or an ecotourist's prize shot, the leopard earns money for local people. The collapse of the fur trade, meanwhile, has significantly reduced poaching.

When you think about it, dumping an animal in an already occupied territory is unlikely to succeed. The new animal is bound to come into conflict with the established residents, who have the advantage of being on home ground and are likely to evict the intruder.

The leopard is an important trophy animal and to use problem animals as such may make more sense than ill-advised translocation. Difficult as it may be to imagine anybody wanting to shoot such a magnificent animal for the pleasure of a trophy, there are plenty of people who do precisely that and are willing to pay for the privilege. To off-set the damage caused by leopards, sustainable hunting of leopards is an option that should not be rejected.

A spotted hyaena (left) caches the carcass of a domestic dog, which is not its usual food. Domestic dogs are a problem in protected areas as they can spread rabies and canine distemper.

THE HYAENAS

Lower Risk

No hyaena species are 'Endangered' or even 'Vulnerable'; all four species have an IUCN classification of 'Lower Risk'. However, some populations, especially of striped hyaena, are becoming ever more fragmented and, in many localities, extinct. Such fragmentation and the limited distribution of the brown hyaena place it and the striped hyaena in the 'Lower Risk' sub-category of 'Near Threatened'.

The most active hyaena predator – the spotted hyaena – is the one most often responsible for killing domestic livestock and, like the lion, has the best chance of surviving in protected areas, where it can reach high numbers. It is in the 'Conservation Dependent' sub-category of 'Lower Risk'.

The aardwolf enjoys the relatively comfortable status of 'Least Concern' in the 'Lower Risk' category. The aardwolf has wrongly been accused of eating lambs. It is exclusively an insect eater. The brown hyaena as well as the striped hyaena may, at times, prey on small domestic animals such as sheep and goats. However, there is no reason why they cannot coexist with people, particularly in areas where sheep and goats are not kept or can be protected at night. In fact, brown hyaenas show an amazing ability to live near large cities, and they make a good living from human refuse.

Although the hyaena is not a trophy animal, many of the issues discussed in connection with lions, leopards and cheetahs are relevant to hyaenas, especially the spotted hyaena.

On a mixed commercial cattle and wildlife conservancy in southern Zimbabwe, an innovative method of herding cattle has greatly reduced predation, in particular by spotted hyaenas. A 'mob' of 40 cattle of the same age are kept together from the time they are weaned and are allocated to a herder. During the day, they graze where they choose, accompanied by the herder, but at night they are brought back to a central area to sleep. The calves are placed in a simple wire enclosure or kraal for their protection. The herder's tent is positioned close to the kraal and a fire is made on the opposite side so that any intruder is visible and can be chased away. After about six weeks, the kraal is moved to prevent overgrazing and excessive trampling.

This imaginative system cuts down predation as well as stock theft and reduces the cost of fence maintenance. The available forage is better used than if the cattle were confined to a camp. The major drawback is that the cattle have fewer hours per day to feed and there are costs in the form of herders' wages. On the positive side, local communities are provided with an opportunity for employment.

A poacher's victim (below), in a snare meant for an antelope.

SERVAL, CARACAL AND THE JACKALS

Lower risk

Although these predators are classified in the 'Lower Risk' category, they raise some important conservation issues. The serval is a wetland species and wetlands are amongst the most vulnerable of habitats as they are frequently drained to support agriculture and building. Wetland conservation is the key to serval conservation, and that of amphibians and a range of species such as cranes. The approach, thus, should be multi-disciplinary.

The caracal and black-backed jackal are often killed in South Africa and Namibia for preying on small stock. An average of 2 200 caracals was killed annually in control operations in the Karoo region of South Africa between 1931 and 1952. These control operations do not appear to have been successful, and the areas were quickly re-colonized.

A bizarre warning, the pelts of black-backed jackals hang on a farm fence to show jackals that they are not welcome. More enlightened approaches are generally more successful.

Once again, a more enlightened approach towards these smaller carnivores on farmland is needed.

The idea that all predators are bad and should be killed still, unfortunately, prevails in most farming areas. The truth is that the major prey of jackals and caracals are insects as well as hares, dassies and rodents, all of which remove much grazing that could otherwise be eaten by sheep or goats. If a farmer poisons all predators in the area, he may actually end up with a bigger problem when faced with the overpopulation of competitors to his sheep. A healthy ecosystem with all or most of its components in place is a better template than a degraded one when farming livestock.

Before embarking on control measures, farmers should weigh the real cost of losses to predators against the cost of predator control. So often the aim of predator control is to decrease or even to eliminate predator populations; it should be aimed, instead, at avoiding damage. I have touched on a number of ways in which this can happen: suitable protection for livestock at night, the use of herders and guard dogs and aversive taste conditioning.

The support of farmers and of the local people is essential if these measures are to be given a chance. The onus on conservationists is to inform and educate the stake holders and to encourage and support their application of viable methods of control. The solutions to all human-wildlife conflict can be found only in a multi-disciplinary approach and through compromise and negotiation.

These are the challenges in the 21st Century for the conservation of African predators.

A golden jackal run over on a main road bordering a protected area (above). Fences to keep predators in reserves may help (below), but game farmers should be encouraged to take down fences and form conservancies.

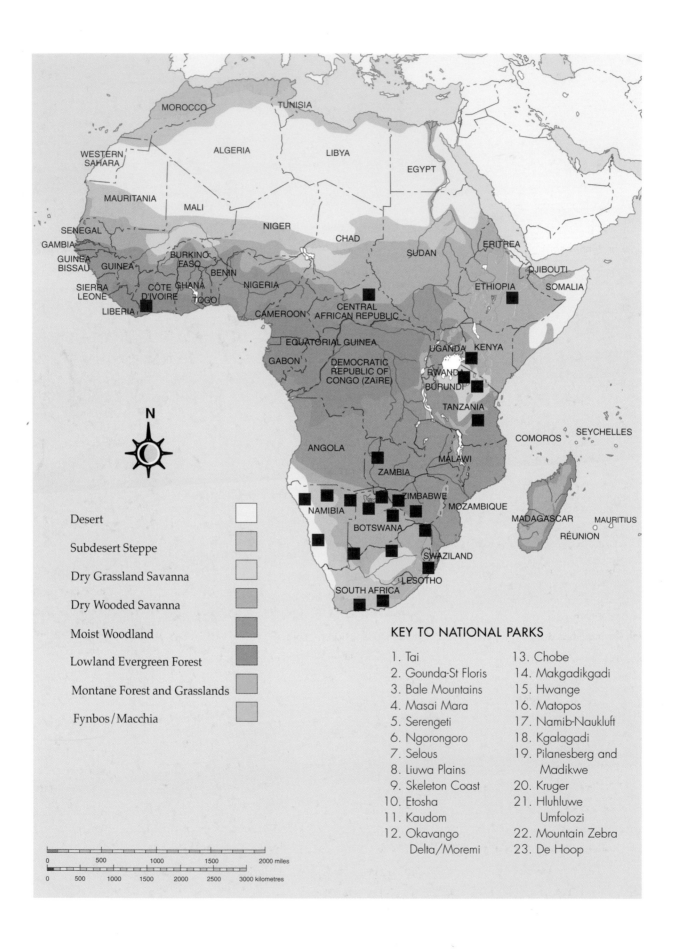

Desert

Subdesert Steppe

Dry Grassland Savanna

Dry Wooded Savanna

Moist Woodland

Lowland Evergreen Forest

Montane Forest and Grasslands

Fynbos/Macchia

KEY TO NATIONAL PARKS

1. Tai
2. Gounda-St Floris
3. Bale Mountains
4. Masai Mara
5. Serengeti
6. Ngorongoro
7. Selous
8. Liuwa Plains
9. Skeleton Coast
10. Etosha
11. Kaudom
12. Okavango Delta/Moremi
13. Chobe
14. Makgadikgadi
15. Hwange
16. Matopos
17. Namib-Naukluft
18. Kgalagadi
19. Pilanesberg and Madikwe
20. Kruger
21. Hluhluwe Umfolozi
22. Mountain Zebra
23. De Hoop

Afro-alpine, African habitat at high altitude, above the tree line, characterized by grass, low shrubs and giant lobelias.

Altruism, type of behaviour for the benefit of others that puts the actor at a disadvantage.

Auditory-bullae, bony projection of the skull housing the structures of the middle and inner ear.

Canidae/canid, a family in the Order Carnivora containing dogs, wolves, jackals and foxes.

Carnivore, member of one of the 20 families of the Order Carvivora.

Clitoris, organ in female mammals corresponding to the penis in males.

Conservancy, collection of farms where wildlife is managed as a single unit.

Copulatory tie, locking of the male's penis in the female's vagina during mating, often seen in dogs.

Cuckold, husband of unfaithful wife (dictionary definition). One that raises an unrelated male's offspring resulting from promiscuous mating (biological definition).

Cursorial, having limbs adapted for running long distances.

Diurnal, active by day.

Ecotourism, low-impact, small-scale nature tourism.

Endemic, occuring only in a particular region, country or island.

Foraging, behaviour associated with looking for and obtaining food.

Home range, the area inhabited by an animal or group that provides all the resources necessary for reproduction and survival over an extended period.

IUCN, International Union for the Conservation of Nature (known as World Conservation Union).

Kin selection, selection of genes resulting from one or more individuals favouring the survival of non-offspring relatives

Kleptoparasitism, stealing of kills of one species by another.

Lactation period, length of time during which a female produces milk.

Land tenure system, manner in which members of a population are distributed in space and time.

Latrine, place where droppings/scats are regularly deposited.

Linear hierarchy, system of dominance from highest to lowest in a social group, where each animal is either dominant or submissive and no two animals have equal rank.

Meta-population, a number of small sub-populations that are managed as a single population.

Miombo woodland, moist savannah found over large parts of Zimbabwe, Mozambique and northwards on acid, usually shallow, soils.

Modal, most frequent or common measurement in a set of measurements that are related.

Multiple paternity, more than one father of a litter.

Natal pack, territory or home range: pack, territory or home range in which an animal was born.

Niche, range of environmental variables such as temperature, humidity, and food items within which a species can exist and reproduce.

Oestrogen, hormone secreted by the ovaries, responsible for changes in a female during the oestrus cycle and for the development and maintenance of female sexual characteristics.

Oestrus, period of heat, or maximum sexual receptiveness, in the female.

Predator, carnivorous animal.

Pasting, depositing of scent marks onto a grass stalk; unique to hyaenas.

Pelage, hair covering or coat.

Polygamy, the practice among male animals of having more than one mate during one breeding season.

Population, set of organisms belonging to the same species and occupying a clearly delimited space at one time.

Relict population, surviving population, characteristic of an earlier time.

Rut, period of sexual activity among male antelope.

Scapula, shoulder blade.

Scat, faeces, droppings.

Sexual dimorphism, physical differences, other than of the sexual organs, often in terms of size, between males and females of the same species.

Sibling or sib, brother or sister, offspring of same parents.

Species, group of interbreeding individuals of common ancestry, reproductively isolated from other groups.

Taxonomy, the science of classification of living things.

Territory, that part of the home range that is defended against other members of the same species.

Testosterone, male sex hormone or androgen responsible for most male physical characteristics.

INDEX